U0444906

塔勒布
智慧箴言录

[美]
纳西姆·尼古拉斯·塔勒布
（Nassim Nicholas Taleb）
著

严冬冬 译

图书在版编目（CIP）数据

塔勒布智慧箴言录：汉英对照 /（美）纳西姆·尼古拉斯·塔勒布著；严冬冬译. -- 北京：中信出版社，2022.9（2022.11重印）

书名原文：The Bed of Procrustes
ISBN 978-7-5217-4487-3

Ⅰ.①塔… Ⅱ.①纳… ②严… Ⅲ.①人生哲学－通俗读物－汉、英 Ⅳ.① B821-49

中国版本图书馆 CIP 数据核字 (2022) 第 115777 号

The Bed of Procrustes: Philosophical and Practical Aphorisms by Nassim Nicholas Taleb
Copyright © 2010 by Nassim Nicholas Taleb
Simplified Chinese translation copyright © 2022 by CITIC Press Corporation
ALL RIGHTS RESERVED
本书仅限中国大陆地区发行销售

塔勒布智慧箴言录
著者： ［美］纳西姆·尼古拉斯·塔勒布
译者： 严冬冬
出版发行：中信出版集团股份有限公司
（北京市朝阳区惠新东街甲 4 号富盛大厦 2 座 邮编 100029）
承印者： 宝蕾元仁浩（天津）印刷有限公司

开本：880mm×1230mm 1/16　　印张：17.75　　字数：100 千字
版次：2022 年 9 月第 1 版　　印次：2022 年 11 月第 2 次印刷
京权图字：01-2011-1721　　书号：ISBN 978-7-5217-4487-3
定价：68.00 元

版权所有·侵权必究
如有印刷、装订问题，本公司负责调换。
服务热线：400-600-8099
投稿邮箱：author@citicpub.com

献给亚历山大·N.塔勒布

普罗克拉斯提斯

在希腊神话中，普罗克拉斯提斯是阿提卡的克雷达勒斯的一个小庄园的主人，这个庄园位于雅典和举行神秘仪式的厄琉西斯城之间。普罗克拉斯提斯，原名达玛斯蒂斯，或波利斐蒙，普罗克拉斯提斯是他的绰号，意思是"拉伸者"。他招待客人的方式非常独特：先引诱旅人进门，为他们提供丰盛的大餐，然后邀请他们在一张非常特别的床上过夜。他要求客人的身高与床的长度必须分毫不差。如果客人个子太高，他就用锋利的斧子把他们的腿截短；如果客人个子太矮，他就把他们的身体拉长。

普罗克拉斯提斯最后可以说是自食其果。他招待的一位旅人正是无所畏惧的英雄忒修斯（他后来在克里特迷宫斩杀了人身牛头怪）。在吃过例行的丰盛大餐之后，忒修斯让普罗克拉斯提斯躺在他自己的床上。为了让普罗克拉斯提斯符合他自己惯常的标准，忒修斯斩下了他的头颅，就像传奇英雄赫拉克勒斯"以其人之道还治其人之身"的行事方式一样。

在更加邪恶的版本［例如伪阿波罗多洛斯的《神话集成》（*Bibliotheca*）中的版本］中，普罗克拉斯提斯拥有两张床，一张大床，一张小床。他让矮个子的客人躺在大床上，让高个子的客人躺在小床上。

这本书中的所有格言在某种意义上都跟"普罗克拉斯提斯之床"有关——遇到我们不了解、不清楚的东西时，我们的解决办法是对我们自己的人生观和世界观进行"拉伸"或者"压缩"，强迫它们符合世俗的、预设的、人为制定的观念、门类和套路。而这会产生爆炸性的后果。我们自己似乎意识不到，这种削足适履的做法，就好像裁缝为了让自己做的衣服刚好符合顾客的身材，而拉长或者截短他们的手脚。例如，几乎没有人会意识到，我们宁可用药物改变孩子的大脑，好让他们跟上学校的课程，也不愿调整课程安排来使其符合孩子的天性。

因为格言一旦解释过多就会失去魅力，所以我点到为止，只在这里提示一下本书的主旨，更进一步的讨论就留到后记中进行吧。这些格言每一条都是独立的，但其内容都围绕着同一个中心思想：在面对我们不了解的东西时，我们是怎样应对的，又应该怎样应对。在我的另外两本书《黑天鹅》和《随机漫步的傻瓜》里，我更加深入地探讨了这一主题。*

---

* 我借用"普罗克拉斯提斯之床"这个典故，并不仅仅为了描述对事物的错误分类，更是为了突出选择错误的变量加以更改的行为——就像更改人的身长而不是床的尺寸。注意，我们所谓的"智慧"（伴以技术层面的娴熟）的每一次失败，都可以归结为"普罗克拉斯提斯之床"式的情况。

目录

前言　V

逆向叙述　001
本体论　009
神圣与凡俗　013
机遇、成功、幸福与坚忍　019
迷人的和不那么迷人的愚人问题　029
像古代人一样生活　035
文字共和国　043
普遍与特殊　053
被随机性愚弄　057
美学　063
伦理　069

- 强大与脆弱　077
- 游戏谬误和场地依赖　083
- 认识论和减法知识　089
- 预言的丑闻　095
- 成为哲学家，并为之继续努力　099
- 经济生活及其他庸俗议题　105
- 贤者、弱者和伟大者　113
- 隐含的和明确的　119
- 爱与非爱　125

- 结尾　129
- 后记　131
- 致谢　137

前 言

你最害怕违拗的人乃是你自己。

当你害怕由一个念头引出符合逻辑的结论时,这个念头就开始变得有意思了。

制药公司更擅长发明和已有药物匹配的疾病,而不是发明治疗已有疾病的药物。

要理解禁欲主义给予人的自由,只要想想看:失去全部财富的痛苦远比失去一半财富的痛苦要轻。

要让傻瓜破产，给他信息即可。

学术与知识的关系，就像卖淫与爱情的关系一样：表面上看似乎很接近，但在旁观者看来，其实并不是一回事。*

科学需要你了解世界，商业需要你让别人误解世界。

我怀疑他们之所以处死苏格拉底，是因为太过清晰地思考是一件非常不讨人喜欢、非常令人陌生、非常违背人性的事。

教育只能让聪明人变得比之前聪明一点儿，却会让傻瓜变得比之前危险很多。

彰显一个创意的独特性的最好证据，不是过去没有跟它类似的创意，而是现在有很多跟它相左的创意。

---

* 我需要澄清的是，这方面有时也有例外，就好像妓女有时也会爱上嫖客一样。

现代社会给我们的双重惩罚是，既让我们衰老得更早，又让我们活得更长。

博学多才的人知道得多，表现得少；记者或咨询人士正好相反。

你的脑子最聪明的时候，是你不告诉它该做什么的时候——人们偶尔会在洗澡时发现这一点。

如果你的愤怒随着时间的推移逐渐消散，说明你对别人做了不公平的事；如果它随着时间的推移逐渐增加，说明别人对你做了不公平的事。

我不知道，究竟是那些主张"慷慨带来回报"的人意识不到这话的自相矛盾之处，还是他们所谓的"慷慨"只不过是一种巧妙的投机方式。*

---

* 慷慨之举应该是不追求任何回报的，无论是在经济上、社会上还是在感情上。这样的举动应该是义务性的（无条件地履行责任），而不是功利性的（追求集体的——甚至是个人的——收获与回报）。"慷慨"的施予会让施予者感到内心温暖，甚至得到救赎，这没有问题；然而，这样的举动不应该跟纯粹出自责任感的义务性举动相混淆。

那些认为宗教是一种"信仰"的人既不了解宗教，也不了解信仰。

工作会摧毁你的灵魂，它会在你"不工作"的时候偷偷侵入你的脑海，所以一定要慎重选择职业。

在自然状态下，我们永远不会重复同样的事情；在受到禁锢的状态下（在办公室、健身房、上下班路上，以及在进行体育运动时），生活只不过是重复性应力损伤，没有任何随机性可言。

把别人缺乏常识当成借口，这本身就是缺乏常识的表现。

用狭隘的（亚里士多德式的）逻辑约束自己，跟避免致命的逻辑矛盾，并不是一回事。

经济学弄不清楚的是，群体（或者集体）行为远比个体行为更难以预料。

在将动物园的动物与野外的动物进行比较时，不要谈论寿命、安全感和舒适度方面的"进步"。

要是早晨起床时，你就能预测这一天会是什么样子，那么你已经开始靠近死亡了——预测得越准确，你离死亡就越近。

冰与水之间并没有中间状态，但是生与死之间确实有中间状态：雇佣状态。

当你害怕的东西绝大多数都带有令人心痒的冒险性时，你所过的一定是被校准过的生活。

拖延是灵魂对羁绊的抗争。

没人愿意被一眼看透，无论是被别人，还是被自己。

逆向

叙述

对撒谎者最好的报复，就是让他相信你真的信了他的谎言。

当打算做一件潜意识里知道注定要失败的事情时，我们就会征询别人的建议，这样就可以把失败怪罪到别人头上。

当真心想说"不"的时候，你就会更难说出口。

如果说"不"是认真的，你就用不着说第二次。

对你名誉损害最大的是你为了维护它而说的话。

当一个人开始谈论衰老的时候,他就真的开始老了。这是关于衰老的唯一客观的定义。

他们会羡慕你的成功、你的财富、你的聪明、你的相貌、你的地位,但是很少有人会羡慕你的智慧。

人们所谓的"谦逊",其实大多数都是掩饰得比较成功的傲慢。

如果你想让人们读某一本书,告诉他们它名不副实就可以了。

只有当他们开始对你展开人身攻击时,你才算赢得了一场争论。

没有什么比"临时"的安排、赤字、休战和情感关系更加恒久了,没有什么比"恒久"的这些东西更加临时了。

让我们最痛苦的,不是跟没意思的人在一起,而是跟努力表现得有意思的没意思的人在一起。

恨是某一行代码出了错的爱，这错误可以改正，但很难被发现。

假如我的某个死敌发现我恨的是另一个人，我不知道他会不会感到嫉妒。

失败者的特点是会抱怨人类的缺陷、偏见、自相矛盾和缺乏理智，但又不利用这些东西追求自己的欢乐和利益。

你是否真的喜欢一本书，判断标准是你是否会重读它（以及重读了多少遍）；你是否真的喜欢一个人的陪伴，判断标准是你是否愿意再次遇见他——剩下的都是空话，或者是那种现在被称为"自尊"的情绪。

我们会问"他为什么富有（或贫穷）"，而不是"他为什么不更富有（或更贫穷）"；我们会问"为什么危机如此严重"，而不是"为什么危机不更加严重"。

恨远比爱更难伪造。你听说过虚假的爱，但还没听说过虚假的恨。

男子气概的反义词不是怯懦，而是科技。

一般来说，所谓的"好的倾听者"其实对他们倾听的内容漠不关心，只不过他们很擅长掩饰这种漠不关心。

正是人们表现出来的自相矛盾让他们富有魅力。

你记得住自己发出去却没收到回复的邮件，却记不住自己收到了而没有回复的邮件。

人们会把恭维之词留给那些对他们的骄傲感没有威胁的人；至于那些威胁到他们骄傲感的人，他们会用"骄傲"来评价。

从古罗马的老加图开始，人们一直通过斥责下一代人的"浅薄"、赞扬上一代人的"价值"来表现自己的成熟。

要忍住不给别人提出锻炼和保健方面的建议，简直跟自己坚持锻炼一样难。

当表扬一个人没有缺点的时候，你也在指出他没有优点。

当她嚷嚷着无法原谅你做的事情的时候，她已经开始原谅你了。

只有当你很容易感到厌倦的时候，缺乏想象力才算是个问题。

我们将那些把自己当成世界中心的人称为自恋者；将那些把自己和另一个人当成世界中心的人称为恋人。

友谊从来都不是可以被宣告结束的，如果是这样，那么其中至少有一个人是愚人。

绝大多数人害怕失去视听刺激，因为当自己去思考和想象时，他们总是在重复同样的内容。

没有回报的恨远比没有回报的爱更让人显得渺小。你无法根据对方的回应做出反应。

对有同情心的人来说，用新的悲哀替换旧的悲哀，远比用快乐替换悲哀要容易。

年轻人的智慧跟老人的轻佻一样不讨人喜欢。

有些人只有在试图表现得严肃的时候才显得滑稽。

在谈话时，要忍住不把秘密说出来是很难的，就仿佛信息具有生存的欲望和繁殖的能力。

本体论

把未被发现的东西当成不存在的东西，这是近来开始流行的一种疾病。但有些人患有更严重的疾病，他们将未被发现的东西当成不可被发现的东西。

让科学解释生活和生命，就像让语法专家解释诗歌一样。

只有当你可以自由地做任何事情，不需要明显的目的、理由和别人的要求时，你才算真正存在于这世上。

神圣与凡俗

你没法用凡俗的语言来解释神圣的东西，但你可以用神圣的语言来讨论凡俗的东西。

如果你没法直接（不加分析地）辨认出神圣与凡俗之间的区别，那么你永远不会知道宗教意味着什么。你也永远不会理解我们通常所说的艺术。你没法理解任何东西。

过去，人们在工作日穿普通的衣服，在星期天祷告的时候换上正装，今天则正好相反。

为了彰显神圣与凡俗之间的区别，每当跟咨询师、经济学家、哈佛商学院教授、记者之类的人发生任何接触（甚至包括邮件联系）之后，我都会仪式性地沐浴，这会让我感到自己得到了净化，清除了凡俗的玷染，直到下一次接触这些人。

还没有被凡俗玷染的媒体就只剩下书籍了，你看到的其他媒体都试图通过广告来操控你。*

你可以用真话来替代谎言，但是神话只能用故事来替代。

神圣的东西都是无条件的，凡俗的东西都是有条件的。**

历史上各种悲剧的源泉，不外混淆了别人的"无条件"与"有条件"。

---

* 在长时间远离媒体后，我开始意识到，没有哪种媒体不是在（笨拙地）试图卖给你什么东西。我只信任图书馆。拥有一本书并不会彰显人性的弱点，不会迎合我们的表现欲，让我们显得高人一等；书籍之外的商业运作才会腐蚀我们。
** 例如，许多被誉为"无法贿赂"的人，只不过是贿赂起来太过昂贵而已。

饭店用食物吸引你，目的是卖给你酒；宗教用信仰吸引你，目的是卖给你规矩（例如，避免负债）。人们可以理解"神"这个概念，但是不能理解未被说明的规矩、禁令和启示。

我很确定，禁食比节食更容易。如果每次"只是吃一点点"火腿，你就不能算是遵守了宗教的饮食规矩。

要彻底戒掉报纸，只需要连续一年每天阅读上个星期的报纸。

机遇、成功、幸福与坚忍

成功就是在中年时成为你在少年时梦想成为的那个人，其他的都是失控导致的结果。

成功的对立面不是失败，而是把认识的名人的名字时时挂在嘴边炫耀。

现代人需要弄清楚，"富有"和"变得富有"在数学层面、个人层面、社会层面和伦理层面上都不是一回事。

要获得彻底的自由，你不仅需要避免成为奴隶，还需要避免成为奴隶主。*

命运惩罚贪婪者的方式是让他贫穷，惩罚特别贪婪者的方式是让他富有。

绝大多数人自杀是因为羞耻，或者是因为经济与社会地位的丧失，不是因为被确诊患了绝症。

"财富"是一个没有意义的词，因为它没法准确衡量；不如换用"缺乏财富"这个词，意思是在任一时刻，你所拥有的和你想要拥有的之间的差距。

年长的人最美丽的时候，是他们拥有了年轻人所缺乏的东西的时候：雍容、博学、智慧、经验以及波澜不惊的平静。

我去参加一场幸福研讨会，结果发现与会者看上去都很不幸福。

---

* 　　历史上，许多人都曾反复验证和揭示过这一点，最近一次有说服力的论证是由蒙田提出的。

傻瓜们所谓的"浪费时间"往往是最好的投资。

衰老一开始是用记忆替换梦想,到最后是用记忆替换别的记忆。

你总想在不招人羡慕或是妒忌的前提下,避免招人讨厌。

不要读最近 100 年写的书,不要吃最近 1 000 年培植的水果,不要喝最近 4 000 年发明的饮料(只喝水和红酒就够了),不要跟 40 岁以上的普通人说话。普通人从 30 岁就开始衰老和死亡了。

有些职业在外人看来很有意思,其实很无聊。他们说,就连海盗也是这样。

卡尔·马克思(一个有远见的人)发现,要更好地控制一个奴隶,你可以说服他相信他其实是个雇员。

在过去,天主教国家和今天相比有更多的接力赛式的一夫一妻制[①]婚姻，但是人们用不着考虑离婚——那时人们的寿命比今天短，那时婚姻的寿命比今天长。

要想一下子变得富有，可以跟穷人打交道；要想一下子变得贫穷，可以跟富人打交道。

等到你可以长时间什么都不做，什么都不学习，什么都不改进，而又完全没有负罪感时，你就算变得文明了。

如果有人跟你说"我很忙"，他要么是在宣称自己的无能（以及对自己的生活缺乏控制），要么是在试图摆脱你。

罗马帝国和奥斯曼帝国时代的奴隶跟今天的雇员之间唯一的不同，就是那时的奴隶用不着奉承主人。

只有当拒绝收下一笔钱比收下这笔钱让你感觉更好时，你才算是

[①] 接力赛式的一夫一妻制（serial monogamy）指一生中有多名配偶，但同一时间不会有超过一名配偶的生活方式。——编者注

富有。

对于绝大多数人，成功就是从憎恨别人的阵营转换到被憎恨的阵营。

要弄清楚你是否真的喜欢自己现在的位置，摆脱依赖，看看你回来时是否跟离开时一样开心。

爱和幸福之间的区别在于，谈论爱的人通常正处于爱河之中，谈论幸福的人通常并不幸福。

现代性就是创造出没有英雄气概的年轻人、没有智慧的老人，以及没有光彩的生活。

你可以通过问一个人他觉得谁有趣来判断他有多无趣。

对于渴望得到别人关注的人来说，网络是个不健康的地方。

不知道有没有人测量过，在聚会上，一个有点儿成功的、在哈佛读过书的陌生人要花多长时间才能让别人知道他的这些辉煌经历。

人们总喜欢关注"榜样"，其实更应该关注的是"反榜样"——你长大后不想变成的样子。

经常道歉是一个很好的习惯，除了在你真正做错事的时候。

对效率的追求，是我们过上充满诗意的、高贵的、优雅的、富有活力的、英雄主义的生活的主要障碍。

有些人，例如绝大多数银行家，是如此不适应成功，以至他们看上去就像穿着巨人衣服的矮子。

不要太过大声地抱怨别人对你的不公，你可能会提醒那些缺乏想象力的敌人该怎么做。

绝大多数人越是试图摆脱执念，就越会深陷其中。

改变别人的意见就跟改变他们的品位一样难。

我曾在被视为"丑陋"的地方度过最美好的时光，也曾在被视为"风景秀丽"的地方度过最无聊的时光。

健壮当然表示一个人有力量，但除去自然刺激因素，努力健身也表明一个人身上有无法治愈的弱点。

魅力就是侮辱别人又不让他们生气的能力。

那些不认为雇佣就是系统化的奴役的人，要么是看不见，要么正在被雇用。

他们生下来被放在盒子里，回到家住在盒子里，学习时勾选一个个的"盒子"，乘着盒子去上班，工作时坐在盒子里，开着盒子去超市买装在盒子里的食物，乘着盒子去健身房，谈论跳出思维的"盒子"，死后被放在盒子里。这些都是符合欧几里得定义的、四四方方的、表面平滑的盒子。

现代性的另一个定义：谈话越来越能通过同一时刻地球上发生的其他谈话的片段被完整地重建。

20世纪宣告了社会乌托邦的破产，21世纪将会宣告科技乌托邦的破产。

建设社会、政治和医疗乌托邦的努力引发了不少可怕的后果，许多科技和医学发明都源于战争。

网络让人们"彼此相连"，在信息和伪社会的层面上营造出了一种怪异的滥交氛围，这让人每次下线的时候都感觉自己重新变得干净了。

在绝大多数争论中，人们似乎都在试图说服对方，但他们其实最多只能说服自己。

迷人的和

不那么迷人的

愚人问题

对于那些你偷偷观察的在饭馆里争吵的夫妇来说，生活中最令人沮丧的一面是，他们几乎永远意识不到自己争吵的真正主题是什么。

给别人提出最多建议的，正是那些最不成功的人，特别是在写作和财务方面。

流言只有在被否认的时候才有价值。

长期来看，你更容易愚弄自己而不是别人。

世上有两类人：追求胜利的人和追求在争论中得胜的人。他们从来不是同一拨人。

人们经常道歉是为了将来还能再道歉。

数学之于知识，好比义肢之于真正的肢体；有些人截肢就是为了能装上义肢。

现代生活会用愚人的方式解释各种活动：现在我们"为了锻炼而散步"，而不是毫无理由地"散步"；我们做事总会有某些隐藏的原因。

社交媒体严重反社会，健康食品非常不健康，知识工作者极度缺乏知识，社会科学根本就不是科学。

对于许多人，我们不应该在他们死去时寻找"死亡原因"，而应该在他们活着时寻找"活着的原因"。

那些利用别人的人，在被人利用时最为恼火。

如果有人给你不止一个理由说明他为什么想得到这份工作，不要雇用他。

二阶思维的失败之处：他告诉你一个秘密，想让你保守这个秘密，而他的行为刚好证明了他自己都不能保守它。

社交网站会注明人们喜欢哪些东西，然而，如果注明他们讨厌哪些东西，它提供的信息就会更有用。

人们太过重视因果逻辑，所以，如果你在谈话中不时插入一句"为什么"，你就可以让最沉默寡言的人变得话多起来。

我得经常提醒自己，真正的独立思考者看上去可能像个会计。

像古代人一样生活

让人上瘾的东西，害处最大的三种是海洛因、碳水化合物和月薪。

我衡量成功的唯一标准，就是你有多少时间需要打发。

我不知道狮子（或者食人族）会不会为自由放养着长大的人类付更高的价钱。

如果你走路时非得听音乐不可，那就不要走路，也不要听音乐。

人们在战争中摧毁彼此，在和平时摧毁自己。

体育让男人变得像女人，让女人变得像男人。

科技可以让愚人的生活在各个方面出现退化（甚至陷入危机），同时又让他相信自己的生活正在变得更"有效率"。

科技和奴隶制的区别在于，奴隶很清楚自己并不自由。

只有在任何领域都不跟任何人攀比，你才算拥有真正的生活。

在面对绝症时，自然规律让你忍受短暂的痛苦之后死去，医学则让你忍受漫长的痛苦之后才慢慢死去。

我们对天然的（或老旧的）东西感到满意，例如美丽的风景或者经典油画，却没法对技术感到满意，所以总是强调新版本里微小的改进，对所谓的"2.0版"无比痴迷。

只有在近代,"努力"才变成一件让人骄傲的事情,而不是一件因为缺乏天赋、才能和气度才不得不做的、可耻的事情。

他们觉得有规律的生活是工作六天,休息一天;我觉得有规律的生活是工作一天（或者半天）,休息六天。

他们所谓的"玩"（健身、旅游、运动）看起来更像工作;他们越是努力,就越会深陷其中。

绝大多数现代提高效率的方法都是延期执行的惩罚。

我们是猎人;我们只有在随机应变的时候才真正活着;没有日程安排,只有来自周围环境的惊喜和刺激。

虽然要受到礼貌的约束,但无论做任何事情,你都可以用无聊程度代替时钟,把它当作生物钟。

对于绝大多数人，腐化开始于离开自由、纯净、有丰富社交活动的校园，开始独自被职业生涯和小家庭所困的时候。

对古典主义者来说，竞技运动员是最可悲的：他努力想成为一种动物而不是一个人，但他永远不可能像猎豹一样快，或像公牛一样强壮。

可以迁移的技能：街头斗殴、荒野徒步、勾引异性、涉猎广博。无法迁移的技能：学习、游戏、体育、做实验——总之，就是那些被简化的、有序的东西。

只有当你的谈话（或者文字）很难用别的谈话片段重新构建时，你才算完整地存在着。

英国有时也会出现地中海式的天气，但英国人还是要到西班牙去度假，因为他们不能自由安排自己的时间。

对绝大多数人来说，工作和与其相关的东西会造成慢性损伤。

技术在无影无形的时候最有效。

真实生活和现代生活的区别，跟谈话和两个人分别背诵台词的区别一样大。

当看到别人在跑步机上奔跑时，我总会想到狮群里最强壮的狮子消耗的能量最少，它每天可以睡 20 个小时，别的狮子会替它狩猎。这就像"恺撒建了一座桥"（Caesar pontem fecit）。*

任何非面对面进行的社交活动都会损害健康。

---

*　"Caesar pontem fecit"这句拉丁文的字面意思是"恺撒建了一座桥"，然而其精妙之处在于，它的意思也可以被诠释为"有人替恺撒建了一座桥"。

文字

共和国

写作就是重复自己的话又不让别人发现的艺术。

绝大多数人书写是为了记住某些东西，我书写是为了忘记某些东西。

他们所谓的哲学，我称为文学；他们所谓的文学，我称为新闻；他们所谓的新闻，我称为闲话；他们所谓的闲话，我（宽容地）称为窥淫狂行为。

作家让人记住的是他们最好的作品，政治家让人记住的是他们最糟糕的错误，商人几乎从来不会让人记住。

批评家表面上看似乎是批评作者没有写出他们想要读的书；然而实际上，他们是批评作者写出了他们想写却写不出来的书。

文学与其说是鼓吹某些特质，不如说是粉饰（你的）某些缺陷。

要享受的话，就翻开纳博科夫的作品读一章。要自我惩罚的话，就读两章。

忧郁症并不是文学，就好像励志并不是哲学。

你得随时提醒自己，魅力存在于没有说出来、没有写下来、没有展示出来的东西中。把握沉默需要境界。

只有当一名作家开始教别人写作时，他才算真正失败了。

科学的过程是无聊的，结果是令人激动的；哲学的过程是令人激动的，结果是无聊的；文学的过程和结果都是令人激动的；经济学的过程和结果都是无聊的。

一条好的格言可以让你用不着跟别人展开对话就能一语中的。

就像有的作家喜欢创作,有的作家喜欢"创作过"的感觉一样,有的书你喜欢读,有的书你很高兴已经"读过了"。

天才就是那些缺点比优点更难模仿的人。

对于一般图书,阅读正文,忽略脚注;对于学术类图书,阅读脚注,忽略正文;对于商业类图书,正文和脚注都可以忽略。

一个人越博学,援引别人的话的次数就越少。

失败者评论比他们更加成功的人的作品时,总觉得有必要申明作者"不是"什么——"他不是个天才,但是……","他不是达·芬奇,但是……",而不是先申明作者"是"什么。

你作品里的陈词滥调越多，你的生命力就越弱。

我们所谓的"商业书籍"是一个由书店发明出来的概念，专门用来指那些没有深度、没有风格、没有内涵，也没有语言特色的书。

像诗人和艺术家一样，官僚是天生的，不是后天形成的；正常人需要付出极大的努力，才能把注意力一直集中在如此无聊的任务上。

专业化的代价：建筑师为了给其他建筑师留下深刻印象而建造建筑，模特为了给其他模特留下深刻印象而保持纤瘦的身材，学者为了给其他学者留下深刻印象而写论文，导演为了给其他导演留下深刻印象而不断尝试，画家为了给画商留下深刻印象而创作，然而，为了给出版编辑留下深刻印象而写作的作家一般都会失败。

回应批评家的言论是浪费感情；如果你的作品到他们死后还能经常再版，那就很说明问题了。

当某位作者写道，塔勒布让黑天鹅理论"变得尽人皆知"时，我就知道他打算剽窃我的作品，但是又剽窃得很糟糕。*

习惯读报纸的人面对真正优美的文笔时，就像失聪者进了歌剧院：他们或许会喜欢其中的一两处细节，却在怀疑"这有什么意义"。

有些书的内容没法总结（真正的文学和诗歌），有些书可以被压缩至大约 10 页，大多数书都可以直接压缩至零页。

信息指数增长的时代，就像一个语言失禁的人：他的话越来越多，听他说话的人却越来越少。

如果往深处挖掘，你就会发现我们所谓的小说远比非小说类作品更不像小说，因为其中的想象力通常更匮乏。

---

* 这也表明，他会试图模仿我，表示"我也是这样想的"。

为你读过的书撰写评论，远比为你没读过的书撰写评论要难。

绝大多数所谓的"作家"坚持写作是为了有朝一日找到点儿值得写的东西。

今天的作家通常有两种：要么清楚地描写自己不了解的主题，要么模糊地描写自己不了解的主题。

信息丰富的黑暗时代：公元 2010 年，仅用英语写作的书就有 60 万本出版，其中只能摘录出寥寥几句让人印象深刻的话；相比之下，大约在公元 0 年，新写出来的书只有很少的几本，但其中充满了令人印象深刻的话。

过去，绝大多数人都是愚昧无知的，约千分之一的人有足够的文化，值得与之对话。今天，文盲率大幅下降，但由于社会进步、媒体和经济的影响，只有万分之一的人值得与之对话。

我们更擅长（不自觉地）做出格的事情，而不是（自觉地）进行出格的思考。

愚人很重要的特点就是，意识不到你不喜欢的东西可能有人会喜欢（你今后也有可能会喜欢），反之亦然。

像一个行动主义者那样思考，远不如像一个思考主义者那样行动危险。

文学在掩饰恶行、缺陷、弱点和混乱时会获得生命力，在说教时则会丧失生命力。

普遍与特殊

我靠自己学会的东西，到现在还记得。

普通人会在故事（和情境）中寻找相似之处，更聪明的人则能从中发现差异。

要理解普遍与特殊之间的区别，可以考虑一下，有些人穿得好看一点儿是为了取悦某个人，而不是所有人。

我们会不自觉地放大自己跟朋友的相同之处，跟陌生人的不同之处，跟敌人的相反之处。

许多人是如此缺乏独创精神，以至他们需要去研究历史，寻找哪些错误是自己可以重复的。

人们认为有害的东西，肯定在某些情况下是有益的；人们认为有益的东西，肯定在某些情况下是有害的。系统越复杂，"普遍规律"的适用性就越差。

傻瓜把特例当成惯例，书呆子把惯例当成特例，有些人两样都犯，聪明人两样都不犯。

你想做你自己，表现出独特性；群体（学校、规则、工作、技术）则希望你像被阉割了一般，成为一个普通人。

真正的爱是特殊性打败普遍性，并大获全胜，也是无条件打败有条件，并大获全胜。

被随机性

愚弄

除非能操控周围的环境，否则我们根本无法控制自己去想哪些事、哪些人，就像无法控制自己心脏的跳动一样。

对摩尔定律的补充：人类的群体智慧每 10 年就会降低一半。*

不要消除别人的幻觉，除非你能用另一个幻觉替代它。（但也不必太过努力，新的幻觉用不着比原来的幻觉更令人信服。）

悲剧之处在于，你以为是随机的东西大都在你的掌控之中，反之

---

* 摩尔定律的基本内容是，芯片上集成的元件数量每 18 个月就会翻一番。

亦然。

傻瓜认为自己是特殊的，别人都是普通的；聪明人认为自己是普通的，别人都是特殊的。

医学之所以能愚弄人们如此之久，是因为它的成功都被凸显出来，而它的错误则被掩盖掉了。

愚人的陷阱是，你会关注你知道而别人不知道的事情，而不是反过来。

中世纪的人是一个他不了解的轮盘上的小齿轮，现代人是一台他自以为了解的复杂机器上的小齿轮。

信息时代的灾难在于，信息的害处远比它的益处增长得快。

从老卡图到现代政客*的转变，最能体现出媒体的作用。如果愿意受到惊吓，你可以自己推断一下其作用究竟是什么。

头脑清晰是勇敢行动的结果，而不是反过来。**

绝大多数沉迷于信息—网络—媒体—报纸的人都很难接受，获得智慧的（主要）方法是从头脑中除去垃圾信息。

更优秀的人能容忍别人轻微的反复无常，但不能容忍别人严重的自相矛盾；弱者能容忍别人严重的自相矛盾，但不能容忍别人轻微的反复无常。

随机性很难跟复杂的、未被发现的、无法被探察的秩序区别开来，但秩序本身也很难跟巧妙的随机性区别开来。

---

*   例如美国前阿拉斯加州州长萨拉·帕林。
** 自苏格拉底的时代以来，人们最大的错误就是，认为缺乏清晰的头脑是各种弊病的根源，而不是反过来。

美学

艺术是与未被观察到的事物进行的单向交谈。

本华·曼德博的天才之处就在于，他不必倚仗完美的形式，也能达到美学上的质朴。

美会因不觉得难为情的不规范行为而得到强化，而华丽则会因愚蠢错误的浮华外表而得到强化。

对"进步"的理解：所有我们视作"丑陋"的地方都是人造的、现代的（例如纽瓦克），而不是自然的或历史的（例如罗马）。

我们喜欢不完美——恰如其分的不完美，我们会为原版的艺术品和充满错误的初始版本支付大价钱。

绝大多数人需要听到别人说"这是很美的艺术"才会跟着说"这是很美的艺术"；有些人需要听到两个人或更多的人这样说，他们才会跟着说。

穆泰奈比宣称自己是所有阿拉伯诗人中最伟大的，他是在最伟大的阿拉伯诗歌里这么说的。

机智风趣之所以吸引人，是因为它以不呆板的方式表现出了智慧。

在古典雕塑作品中，男人身材瘦削，女人身材丰腴；在现代摄影作品中，正好相反。

没有哪个学者比哪怕最糟糕的创造者更有价值，就像没有哪只猴子比哪怕最丑的人更漂亮一样。

要是你想惹恼一个诗人，就去诠释他的诗歌。

伦理

要是你需要理由来解释为什么你跟某人是朋友，你们就不算是朋友。

我对现代性最不满的地方，或许就是"合乎伦理"跟"合法"越来越不是一回事。*

生命之美：你生命中遇到的最友善的举动，可能来自不求回报的局外人。**

---

* 前美国财政部长、"银行流氓"罗伯特·鲁宾或许是历史上最大的窃贼，但他并没有犯法。在复杂的社会体系中，"合乎伦理"跟"合法"之间的差异会越来越大……最终让体系垮台。
** 反过来，你遇到的让你最痛苦的事，可能来自在你生命的某些时刻非常关心你的人。

我们最愿意帮助的，偏偏是那些最不需要我们帮助的人。

要衡量一个人的价值，就考虑一下你第一次见到他时的印象，和最近一次见到他时的印象有多大的差别。

冥想是一种不伤害别人的自恋方式。

真正的谦卑，是你能比别人更让自己感到惊讶；剩下的要么是羞涩，要么是巧妙的推销。

如果某个人吹嘘自己的成就，我们会感到恶心；但如果国家吹嘘自己的成就，我们则称为"民族自豪感"。

你只能说服那些认为自己在被说服后可以获益的人。

伟大始于用有礼貌的鄙夷取代仇恨。

那些躺着或站着赚钱谋生的人，比坐着赚钱谋生的人更值得信任。

美德的悲剧在于，一条谚语越是明显、无聊、人云亦云、充满说教意味，就越难以实现。

即使是最小气的人也会慷慨地提供建议。

你要是对我撒谎，那就一直撒下去，不要突然讲真话来伤害我。

不要相信一个需要收入的人——除非他拿的是最低工资。[*]

你或许会因为年老而失去力量，但不会因此而失去智慧。

弱者行动是为了满足自己的需求，强者行动是为了履行自己的义务。

---

[*] 那些为企业所奴役的人会为了"养家糊口"做任何事情。

宗教和道德的演变轨迹：从向你承诺如果做好事就能上天堂，到你做好事时向你承诺有天堂，再到让你承诺做好事。

尽量不要把那些别无选择的人称为英雄。

有些人会因为你给予他们的而感激你，有些人会因为你没有给予他们的而怪罪你。

信守道德的人根据信仰选择职业，而不是根据职业选择信仰。自中世纪以来，这样的人越来越少了。

我信任所有人，除了那些告诉我他们值得信任的人。

人们经常需要暂停自我推销，在生活中找到一个不需要他们去取悦的人，这就解释了为什么许多人喜欢养狗。

帮助忘恩负义的人才是纯粹的慷慨。其他形式的帮助都是为了自己。*

我不知道骗子能不能想明白，诚实的人可能比他们更精明。

普鲁斯特笔下有一个角色名叫莫雷尔，他把借给他钱的犹太人尼西姆·贝尔纳当成魔鬼，甚至变成一名反犹太主义者，只是为了摆脱感激之情。

祝福做了好事的人会交好运，这听起来像是行贿，或许这是古老的、前道义主义和前古典主义道德观的残余。

伟大跟骄傲的区别，体现在没人看的时候你会做什么。

民族国家：没有政治错误的种族隔离。

---

*　　这是康德伦理学的观点。

在 100 个人中，50% 的财富、90% 的想象力、100% 的智慧都集中在一个人身上——不一定是同一个人。

就像染头发会让老男人失去魅力一样，你越是掩饰自己的弱点，它们就会越令人反感。

对于士兵，我们有"雇佣兵"①这种说法，但对于雇员，我们却说"人人都需要谋生"。

有些人的骄傲表现为对强者的不屑，有些人的骄傲表现为对小人物的蔑视。

如果你所在的社会阶层里有一个人变穷了，那么这种事会比别的社会阶层里成千上万的人变穷了更能触动你。

---

① 原文用了"mercenary"一词，它除了表示雇佣兵，也表示人唯利是图。——编者注

强大与脆弱

当你可以失去全部财产,又不用为此而变得谦卑时,你才算是安全的。*

要测试一个人在承受名誉损毁方面有多强大,可以当众问他"是不是还做得不好"或者"是不是还在损失金钱",然后看他的反应。

稳健是不急不躁地进步。

---

*　这是我曾曾曾曾曾祖父提出来的。

如果在两种选择之间摇摆不定，那就两种都别选。

民族国家喜欢战争，城邦喜欢商业，家庭喜欢稳定，个人喜欢娱乐。

当你在乎喜欢你作品的少数人甚于不喜欢你作品的多数人时（艺术家），你是强大的；当你在乎不喜欢你作品的少数人甚于喜欢你作品的多数人时（政客），你是脆弱的。

理性主义者想象着没有傻瓜的社会；经验主义者想象着不受傻瓜影响的社会，或者更棒的，不受理性主义者影响的社会。

学者只有在试图成为无用的人时才有用（例如在数学和哲学中），而在成为有用的人时则会很危险。

对于强大的人，错误提供了信息；对于脆弱的人，错误就是错误。

测试你对名誉损失的抵抗力的最好方法就是，想想你收到一名记者的邮件时会有什么情绪反应（害怕、开心，还是无聊）。

成为一名作家的主要不便之处是，你公开或者私下做的任何事情都不会损害你的名誉，特别是在英国。

当（国家和个人的）仇恨的对象改变时，原先强烈的仇恨就结束了；一般人没法同时应对一个以上的敌人。正因如此，由时而结成同盟、时而互相征战的城邦组成的系统才是一个强大的系统。

我发现，讨厌大政府却喜欢大企业是一件不搭调（且有问题）的事，但反过来则不是。

你乘坐的跨大西洋航班晚点一个小时、三个小时、六个小时的频率分别是多少？提前一个小时、三个小时、六个小时到达的频率又分别是多少？这解释了为什么缺陷的程度总是比预计的更严重，而不是更轻微。

游戏谬误
和
场地依赖*

* 《黑天鹅》中揭示的一条谬误是,让生活像游戏一样充满了刻板的规矩,而不是反过来。场地依赖是指,一个人在一种环境下(例如在某个体育场里)采取某种行为方式,换一种环境就采取另外的行为方式。

体育是商品化甚至卖淫化了的随机性。

当你在肉体上战胜一个人时,你得到了锻炼,缓解了压力;当你在网上攻击他时,你不过是在损害自己。

平整的地面、竞技体育和专项工作会使头脑和身体僵化,竞争激烈的学术环境则会使灵魂僵化。

他们承认象棋训练只能增强下象棋的技能,却不承认课堂训练(几乎)只能增强课堂技能。

一到达迪拜的酒店，那个商人就让服务员替他搬行李，后来我看见他在健身房里做举重锻炼。

比赛被创造出来是为了让普通人产生胜利的幻觉。在现实生活中，你不知道谁真正赢了，谁真正输了（除非为时已晚），但你知道谁很英勇，谁不英勇。

我怀疑智商测验、美国高考和学业成绩是书呆子发明出来的，这样他们可以得到高分，显示自己的聪明。*

他们可以在电子阅读器上读吉本的《罗马帝国衰亡史》，却拒绝用泡沫塑料杯喝靓茨伯庄园的红酒。

关于思维的场地依赖，我能举出的最好的例子是，前段时间我去巴黎，午餐在法式餐厅吃，我的朋友们吃了鲑鱼肉，把鱼皮扔掉了；晚餐在日式寿司餐厅吃，同样是这帮朋友，他们吃了鲑鱼

---

* 有些聪明又睿智的人在智商测验中得了很低的分数，也有些智力明显有缺陷的人（例如美国前总统乔治·w. 布什）在智商测验中得了很高的分数（130）。与其说是他们接受了智商测验，不如说是他们测验了智商测验本身。

皮，把鱼肉扔掉了。

脆弱：我们一直在努力把勇气与战争区分开来，让会用计算机的胆小鬼不必冒一点儿生命危险就可以杀人。

认识论

和

减法知识

自柏拉图以来，西方人的思想和知识理论一直把真伪矛盾作为关注的焦点。尽管这很不错，但我们确实该把关注点转移到强大与脆弱的矛盾，以及愚人与非愚人的社会认识论上了。

知识的问题在于，由观鸟者撰写的关于鸟类的书籍，远比由鸟类撰写的关于鸟类的书籍和由鸟类撰写的关于观鸟者的书籍多得多。

完美的愚人知道猪会盯着珍珠看，却不知道自己有时也处在类似的情境中。

我们需要杰出的智慧和自控力才能承认，许多事情的逻辑是我们所不了解的，并且比我们自己的逻辑更加明智。

知识是我们减掉的内容（什么行不通，什么不该做），不是我们增加的内容（什么该做）。*

他们认为聪明就是注意到有关系的事情（探测模式）；在复杂的世界里，聪明就是忽略无关的事情（避免错误模式）。

幸福：我们不知道它是什么意思，怎样测量它，怎样实现它，但我们非常清楚如何避免"不幸福"。

天才的想象力远远超出他的智力，学者的智力远远超出他的想象力。

理想的、对社会和学生危害最小的三学科教育是教授数学、逻辑学和拉丁文。拉丁文著作的阅读量要加倍，以补偿数学带来的智

---

* 吹牛的人：告诉你该做什么，而不是不该做什么的人（例如咨询师或者股票经纪人）。

慧流失；学习数学和逻辑学，只要能避免说空话和花言巧语就够了。

影响力最大的四个现代人：达尔文、马克思、弗洛伊德和（高产的）爱因斯坦，他们都是研究学问的人，但不是高校科研人员。在院校内做真正有意义的、长久的学问，从来都是件难事。

预言的丑闻

先知不是具有特殊预见力的人，而是对别人看到的绝大多数东西视而不见的人。

对古人来说，预言历史事件是对（诸）神的亵渎；对我来说，这是对人类的亵渎；对有些人来说，这是对科学的亵渎。

古人很清楚，理解事件的唯一方式就是促使它们发生。

任何做出预言或者表达意见而不必承担风险的人，都有些虚伪。除非他会跟着船一起沉下去，否则，对他来说，这就像看一场冒险电影。

要让他们更加严肃地看待预报内容,可以告诉他们,在闪米特人的语言里,"预报"和"预言"是同一个词。

塞涅卡认为,斯多葛派的智者在国家腐败到无可挽回、自己的努力没有效果时就应该退隐。等待国家自我毁灭是更明智的做法。

成为哲学家，

并为之

继续努力

要成为一名哲学家，首先走路要非常慢。

真正的数学家能理解完整性，真正的哲学家能理解不完整性，其余的人理论上什么都不理解。

25个世纪以来，一直没有出现在聪明、深度、优雅、才智和想象力方面堪比柏拉图的人——让我们免受柏拉图的思想遗产的影响。

我为什么一直痴迷于柏拉图的问题？绝大多数人需要超越前人，而柏拉图成功超越了所有后来人。

哲学家经过长时间的散步，仅仅通过推理就能弄清楚事情是什么样的，是先验的；别人只能通过错误、危机、事故和破产来弄清楚事情是什么样的，是后验的。

工程师会计算而不会定义，数学家会定义而不会计算，经济学家既不会定义也不会计算。

有限但上限未知的东西，在认识论上与无限的东西没什么两样。认识具有无限性。

有意识的无知（如果你做得到）可以扩展你的世界，它可以让事物变得无限。

对古典哲学家来说，哲学洞察力是一辈子闲散生活的结果；对我来说，一辈子闲散生活是哲学洞察力的结果。

要承认有道理的东西并不是真的有道理，需要出众的才智和强大的自信心。

神学上的普罗克拉斯提斯之床：对东正教徒而言，自格利高里·帕拉马斯以来，对阿拉伯人而言，自阿尔加惹尔以来，人们一直在试图用哲学上的普适语言来定义上帝，这是一个理性主义的错误。我还在等待某个现代人注意到这一点。

说"不确定性的数学"就像说"性的贞洁"一样——数学化的东西就不是不确定的，反之亦然。

可悲的是，我们从傻瓜、经济学家和别的反面教材身上学到了最多的东西，我们对他们却一点儿都不感激。

在柏拉图的《普罗塔戈拉篇》里，苏格拉底将哲学与诡辩家的花言巧语做了对比，哲学是为了真理而共同探寻，诡辩家使用花言巧语则是为了在关于名誉和金钱的辩论中取得胜利。25个世纪之后的今天，领薪水的研究人员和喜爱终身职位的学者就是如此。这就是所谓的"进步"。

经济生活及其他庸俗议题

有些行当，例如"经济学家"、"妓女"和"咨询师"，即使进一步描述也提供不了更多信息了。

数学家从问题出发得出解决方案，咨询师从"解决方案"出发创造问题。

他们所谓的"风险"，我称为机遇；他们所谓的"低风险"机遇，我称为愚人的问题。

组织就像是受了咖啡因刺激的受骗者，不知不觉地向后倒退，你只会听说其中极少数人到达了终点。

要判断一个人是否极度愚蠢（或者极度聪明），就看经济和政治新闻对他来说是否有意义。

左派认为市场是愚蠢的，所以模型应该是聪明的；右派认为模型是愚蠢的，所以市场应该是聪明的。双方都未意识到，市场和模型都是极度愚蠢的。

经济学就像一颗死亡的恒星，尽管看上去仍然在发光，但你知道它已经死了。

愚人认为你可以用金钱治疗贪婪，用毒品治疗毒瘾，用专家治疗专业问题，用银行家治疗银行业问题，用经济学家治疗经济学问题，用欠债治疗债务危机。

如果一家企业的老总公开宣布"没什么需要担心的"，那么可以肯定，他有很多需要担心的。

简单总结一下股票市场：参与者平静地在即将被宰割的队伍后排

队等待，还以为自己是为了百老汇的表演而排队。

政府救市和抽烟的主要区别在于，在某些罕见情况下，"这是我的最后一根烟"，这句话有可能是真的。

让我们脆弱的是，组织机构不可能拥有跟个人同样的美德（有荣誉感、诚实、勇敢、忠诚、坚忍）。

最糟糕的损害是能力足够的人试图做好事而造成的，最棒的改进是能力不足的人不想做好事而引发的。

银行和黑手党的区别在于：银行更擅长跟法律法规打交道，但是黑手党能理解公众意见。

"骗数十亿美元比骗几百万美元容易多了。"*

---

* 这是某些人对麦道夫欺诈案的看法。

在莫斯科的会场上，我眼看着经济学家埃德蒙·费尔普斯因为写的东西没有人看，提出的理论没有人用，发表的演讲没有人懂，而获得了"诺贝尔奖"。

在非线性领域，"科学估计"的失败之一在于，期望值的平均值跟平均值的期望值并不一样。不要因为一条河"平均"只有四英尺[①]深就试图蹚过去。这一规律又叫詹森不等式。

记者跟格言家的思路正好相反：我提出"你需要技能才能赢得宝马车，需要技能和运气才能变成沃伦·巴菲特"，记者们对此的总结是"塔勒布认为巴菲特不具有任何技能"。

好奇的人喜欢科学，有天赋而敏感的人喜欢艺术，实干的人喜欢商业，剩下的人只好当经济学家了。

上市公司像人体细胞一样，注定会衰亡，会因为债务和潜在风险而走向自我毁灭。救市行为让这一过程有了历史维度。

---

① 1英尺等于0.3048米。——编者注

在贫穷国家，官员可以收下明摆着的贿赂；在华盛顿，他们会得到为大公司工作的、复杂巧妙的、含蓄的、心照不宣的承诺。

当一个银行家陷入穷困时，命运是最残酷的。

我们应该让学生重新计算学分，方法是把金融和经济学方面的得分倒扣回去。

由于潜在风险的积累，代理问题会让每家公司都变得极其脆弱。

在政治里，我们要么选择酷爱战争和民族国家的大企业代表，要么选择盲目冒险的、与高层有关的、在认知上傲慢自大的大雇主服务者，但我们至少还有选择。

贤者、弱者和伟大者[*]

[*] 在亚里士多德的《尼各马可伦理学》中,"伟大者"是指"灵魂伟大"的人,他们认为自己配得上伟大的事物,清楚自己在生活中的位置,遵守排除一切偏狭和卑劣的伦理体系。"灵魂伟大"的概念尽管被提倡谦卑的基督徒置换掉了,但黎凡特文化仍然保留着它。"伟大者"具有许多特点,其中包括走路很慢。

普通人往往会被小的冒犯激怒，但却在大的冒犯面前保持沉默，无动于衷。*

主导者的唯一定义是：如果你努力成为主导者，那么你永远也成不了主导者。

那些没什么要证明的人，从来不说他们没什么要证明。

弱者展示力量，掩饰弱点；伟大者展示弱点，就像展示饰品一样。

---

* 想想人们对银行业和经济机构的反应。

要是能变得睿智而又不无聊该有多好，变得无聊却又不睿智是多么糟糕。*

让我尊重的特质包括博学，以及在别人顾忌名誉的场合挺身而出的勇气。任何白痴都可以很聪明。

普通人更容易为说出的话而不是沉默后悔，比较杰出的人更容易为沉默而不是说出的话后悔，伟大者没什么可后悔的。**

只要请普通人吃若干顿饭，他就会替华盛顿的美联储撒谎、偷窃、杀人，甚至成为其预报员；伟大者永远不会。

社会科学意味着创造出某种我们能理解的人类。

跟地位相当的人说"祝你好运"时，弱者的想法正好相反，强者

---

*　看看前美国联邦储备委员会主席本·伯南克。

**　我在读了10遍亚里士多德的《尼各马可伦理学》第四卷之后才意识到，他没有明确地说出来（但内心很清楚）的内容是：伟大的东西都是无条件的。

无所谓，只有伟大者才是真心的。

在过去，只有一部分男性有生育权，但是所有的女性都有生育权。平等对女性而言更为自然。

伟大者只相信他所听到的东西的一半，但是双倍相信他自己所说的东西。

口头威胁是能力不足的最好证明。

历史上最著名的两位勇敢者并不是荷马史诗里的战士，而是地中海东部的两个人，他们为自己的观点而死，甚至可以说是故意寻死。

弱者不可能是好人，或者说，只有在太过全面、毫无遗漏的法律系统里，他才能算是好人。

无论如何，尽量避免说话——威胁、抱怨、解释、叙述、重复、恳求、试图赢得胜利的争论、祈求；尽量避免说话！

按照萨莫萨特的琉善的记载，哲学家泽莫纳克斯曾阻止一名斯巴达人殴打他的仆人。"你这样做等于把他变成了跟你对等的人。"他说。

古人最怕的是不光彩的死亡，现代人最怕的是死亡。

隐含的

和

明确的

当人们注意你的缺席胜过别人的在场时，你就知道自己的影响力了。

如果有人宣称"没有下次"，那么他肯定会再做一次。

有些沉默寡言的人用沉默来掩藏自己的智慧，但绝大多数人这样做是为了掩藏自己缺乏智慧的事实。

当有人说"我没那么蠢"时，往往意味着他比自己想象的更蠢。

辱骂是唯一一种绝不可能造假的欣赏方式。

当女人说男人聪明时,她通常是说他长得帅;当男人说女人笨时,他肯定是说她长得漂亮。

婚恋网站没有理解的是,人们对自己没说出口的东西更感兴趣。

你通常宁愿跟那些认为你有趣的人在一起,也不愿跟那些你认为有趣的人在一起。

互联网打破了公私之间的壁垒,过去那些私下里一时冲动的和不雅的表达现在都通过文字呈现出来。

社交网站的一个问题是,别人想在你背后埋怨你变得越来越难了。

如果有人对你说"我做不了别的什么了",那么他肯定可以帮助你,只是他不愿意这么做。如果他说"我来帮你",那么他肯定既没法帮助你，也不愿意这么做。

我们知道景点和商品肯定比宣传手册上描述的差，但我们没法容忍人们表现得比他们留给我们的第一印象差。

如果有人讲话时以"简单地说"开头，那么他很可能会谈论非常复杂的事情。

一半人用嘴巴说谎，另一半人用眼泪说谎。

爱与

非爱

在任何时候，人们都会渴望得到金钱、知识或者爱；有时一个人会同时渴望得到其中的两样，但不可能同时渴望得到这三样。

没有牺牲的爱就像是偷窃。

婚姻是让男人女性化——同时也让女人女性化——的制度化过程。

有些男人让自己被女人（以及财富）包围，是为了炫耀；另一些男人这样做主要是为了消费，两类人基本上不重合。

除了友谊和爱，你很难找到参与的双方都是愚人的情况。

我参加过一次 symposium（专题研讨会），这个词在公元前 4 世纪指的是雅典的酒会，在酒会上，未被书呆子化的雅典人会谈论爱的话题。这次研讨会上没有酒，幸好也没人谈论爱。

最关注你的人是那些恨你的人。朋友、追求者和伴侣都不可能对你如此好奇。

当一个年轻女人跟一个不怎么有趣的富有男人恋爱时，她可以真诚地相信，她是被他的某个特殊的身体部位（例如鼻子、脖子、膝盖）迷住了。

一个好的敌人远比最有价值的追求者更忠诚，更容易预料，并且对聪明人来说也更有用。

如果诽谤我的人更了解我，他们就会更恨我。

结 尾

柏拉图式的思维认为，生活就像电影，必然会有结尾。非柏拉图式的思维认为，电影就像生活，除了某些无法逆转的情况（例如死亡），他们不相信任何人宣称的结尾是真正的终结。

后 记

我作品的普遍主题是人类知识的局限,以及我们在跟自己视野范围之外的事物——未知的、隐身在不透明的帷幕另一边的事物——打交道时,会犯下的令人着迷和没那么令人着迷的错误与偏见。

因为我们的思维需要减少要处理的信息量,所以我们更愿意将事物硬塞进已知框架中,放在普罗克拉斯提斯之床上(对未知事物进行"截肢"),而不是暂停对事物的分类,让事物变得清晰明确。我们发现了越来越多的虚假模式和真实模式,随机的东西会因此变得不那么随机,会越来越有确定性——我们过度活跃的大脑更愿意采用错误的、简化的叙事模式,而不是放弃套用任何叙事模式。*

---

* 这种对未知事物的轻视源于人类"对抽象内容的不屑"(我们的思维不善于处理轶闻趣事以外的内容,而且很容易受到生动的图像的影响,这让媒体得以扭曲我们的世界观)。

思维可以是一种很好的自我欺骗工具——它生来就不适合处理复杂性问题和非线性的不确定性问题。*跟常见的观念正好相反，更多的信息实际上意味着更多的错觉。现代性和信息时代带来的影响之一是，我们能越来越快地发现虚假模式。在今天这个充满信息的复杂世界中，混乱的随机性与我们源自简单的原始生活环境的、对事物的直觉已不匹配。我们的精神架构与我们所生活的世界也越来越不匹配。

这就导致了愚人问题：当地图跟某一片区域的实际情况不相符时，某一类傻瓜——受到过度教育的人、学者、记者、读报纸的人、思维僵化的"科学家"、伪经验主义者、那些具有我所说的"认知的傲慢"（也就是不理会他们没有看到、没有发现的事物）的人——会进入一种否定状态，想象这片区域的实际情况和地图相符。更笼统地说，这里所说的傻瓜是那些为了削减而进行了错误削减的人，或者说他们移除了某些至关重要的东西，他们截掉了客人的腿脚，或者，稍好一点的，砍掉了客人的一部分头颅，却坚持说这样仍旧保留着客人95%的外形。看看我们自己创造的普罗克拉斯提斯之床吧，其中有些是有益的，有些则是有问题的：规章制度、等级分明的政府、学术界、体育馆、上下班路线、高层办公楼、非自愿建立的人际关系、雇佣等等。

自启蒙运动以来，在理性主义（为了让事物有意义，我们希

---

* 科学也不能有效地处理非线性问题和复杂问题，这些问题（涉及气候、经济、人体等）相互依存，尽管科学在线性领域（物理学和工程学）取得了令人兴奋的成功，但是它的声望也给我们带来了危机。

望它们是什么样子的）和经验主义（事物实际的样子）的对立中，我们一直在责怪这个世界不符合"理性"模型的普罗克拉斯提斯之床，一直在尝试改变人类以适应技术，改变伦理以适应我们对雇佣的需求，改变经济生活以适应经济学家的理论，努力把人类的生活塞进某种叙事框架之中。

当我们对未知事物的描绘和对随机效应的理解出错时，如果这些错误不会导致负面结果，我们就是强大的——否则我们就是脆弱的。强大的人能够从黑天鹅事件[*]中获益，脆弱的人则会受到此类事件的严重冲击。我们越来越容易受到某种科学的自我中心主义的影响，对未知的东西提出充满自信的结论，这导致了专家问题、风险，以及对人类错误的严重依赖。读者从我的格言中可以看出，我对大自然维持强大的方法心存敬意（几十亿年足够让任何脆弱的东西土崩瓦解）；古典思想（因其对未知事物的尊重、在认识论上的谦卑）比现代后启蒙时期的幼稚的伪科学的自我中心主义更加强大。因此，我的古典价值观驱使我强调博学、优雅和勇气的重要性，反对现代性中的虚伪、书呆子气和庸俗。[**]

艺术是强大的，科学并不总是强大的（这是较为温和的说法）。某些普罗克拉斯提斯之床使得生活更有意义：艺术，以及

---

[*] 黑天鹅事件是指未被任何人预测而又造成了巨大影响的（历史、经济、科技或个人的）事件。尽管我们的知识在不断增长，但黑天鹅事件发挥的作用也越来越大。
[**] 许多庸俗的人将我的观点简化为反对科技，而事实上，我反对的只是人们那天真的对科技的副作用视而不见的态度——这是脆弱的典型表现。我宁愿无条件地接受道德，有条件地接受科技，也不愿反过来。

所有事物中最强有力的东西——诗歌般的格言。

格言、警句、谚语、俗语，甚至讽喻性的短诗，可以说是最早的文学形式，今天我们通常将它们统称为诗句。它们具有金句的紧凑（尽管它们比今天低端市场版本的金句更强大，也更优雅）*，带着一点点虚张声势的意味，展示了作者把强有力的观点压缩为寥寥几句话——尤其是几句口语的能力。的确，这无疑是一种虚张声势，因为阿拉伯语中表示"即兴创作的俏皮话"的词语本来就有"有男子气概（manliness）的行为"的意思，尽管这里所谓的"男子气概"并不像字面上那样仅仅局限于男性，它更适合翻译为"做人的技能"["美德"（virtue）一词也是这样，发源于 vir 这个拉丁文词根，而 vir 的意思就是"人"]。就好像那些能够用这种方式表达出强有力的思想的人身上有某种魔力。

这种方式是黎凡特文化（以及更广泛的地中海东部文化）的核心。当上帝对闪米特人讲话时，他讲的都是充满诗意的短句，通常借由先知之口说出来。《圣经》，尤其是《圣经·箴言》和《圣经·传道书》，伊斯兰教的圣书《古兰经》都是浓缩的格言集。类似的形式也为后人所借鉴，用于创作文学预言：尼采的《查拉图斯特拉如是说》，以及距现在较近的、来自黎巴嫩北部村庄的我的同胞哈利勒·纪伯伦的《先知》，都采用了格言集的形式。

除了我们今天所谓的宗教，知名的格言还有赫拉克利特和希

---

* 注意格言与电视上的俏皮话的区别：金句会导致信息的流失，而格言会带给人更多的信息。在某些方面，格言遵循吉仁泽与戈尔茨坦提出的"少即是多"效应。

波克拉底的格言，普布利留斯·西鲁斯的格言［他本来是叙利亚的奴隶，因为流利的口才而获得自由，他强有力的一行诗都收录在他的《格言集》(Sententiae)中，法国思想家拉罗什富科的不少格言都受到他的影响］，以及普遍被认为是阿拉伯最伟大诗人的穆泰奈比的诗作。

作为独立的句子，格言曾被用于阐释观点，被当作宗教经文，一位黎凡特祖母将之作为给孙子的建议，它们也曾被用于吹嘘自己（我在之前的一条格言中提到过，穆泰奈比用格言令人信服地告诉我们，他是最伟大的阿拉伯诗人），被用来讽刺*（马提雅尔、伊索、艾尔玛利），被道德家使用（沃夫纳格、拉罗什富科、拉布吕耶尔、尚福尔），被用于解释晦涩的哲学（维特根斯坦）、相对清晰的哲学（叔本华、尼采、萧沆）或十分清晰的观念（帕斯卡）。**你永远不必解释格言的意思——像诗歌一样，这是读者需要自己消化的东西。***

有些格言非常平淡，有些是老生常谈，这些格言中蕴含了你思考过的重要真理（这样的内容会让聪明人在纪伯伦的《先知》面前畏缩）；有些格言令人愉悦，其中的内容你从未思考过，会

---

* 在互联网时代，表明人们文化修养缺失的最佳证明，坦率地说，人们"书呆子化"的最佳证明，就是讽刺日益消失，僵化的思想只能接收到字面意义上的侮辱。

** 同一句格言经常会被不同时代、不同地域的好几位作者重复。

*** 格言（除了德语格言）的质量已经有所下降，因为它总是和奥斯卡·王尔德、马克·吐温、安布罗斯·比尔斯、萨卡·圭特瑞的俏皮话联系在一起。深邃的思想可以是诗意的、风趣的，就像叔本华、尼采或（有些时候）维特根斯坦的话那样；但是，神圣的东西与凡俗的东西是分开的，哲学与诗歌并不是单口相声。

让你情不自禁地发出"原来是这样"的感慨（例如拉罗什富科的格言）；然而，最棒的格言是这样的：其中的内容你从未思考过，并且需要读好几遍才能意识到其中包含的重要真理。有些时候，沉默寡言的品格在这些真理身上是如此突出，以至你读完它们之后马上就忘记了。

格言要求我们改变阅读习惯，每次只读一点点，因为每一句格言都是一个整体，是一段完整的、独立的叙述。

我对书呆子的最好定义是：一个要求你解释格言内容的人。

我清楚我的语言风格是格言式的。十几岁的时候，我接受过诗人乔治·谢哈德的指导（他的诗歌读起来像格言），当时他就预言，将来我会顿悟并投身诗歌领域，只要我把"思想产业"从脑子里赶出去。最近，有读者在网上引用我书里的话，由此引发了大量的版权警告，但是我从来没有想过要用这种方式来重述我的观点（或者，不如说是我关于知识的局限性的核心观点），直到我意识到，这些句子是自然而然地出现在我脑海中的。我几乎是无心地、以一种怪异的方式想出它们的，尤其是在我（缓慢地）散步时，或当我放空思想什么都不做时，或不做任何费力气的事情时，我可以确信，我能听到从不透明的帷幕的另一边传来的声音。

如果你能让自己彻底摆脱所有的限制、所有的想法、所有让你精疲力竭的被称为工作的活动、所有的努力，隐藏在现实纹理中的元素就会开始凸显。接着，你从未思考过的奥秘就会出现在你眼前。

致 谢

感谢 P. 塔诺斯、L. 德尚特、B. 奥贝蒂特、M. 布利斯、N. 瓦迪、B. 爱珀雅德、C. 米海勒斯库、J. 巴兹、B. 杜皮埃、Y. 兹尔伯、S. 罗伯茨、A. 皮尔贝、W. 古德拉德、W. 墨菲、M. 布罗克曼、J. 布罗克曼、C. 塔勒布、C. 桑迪斯、J. 库雅特、T. 波恩海姆、R. 杜贝利、M. 格森（比较年轻的那一位）、S. 塔勒布、D. 里维埃、J. 格雷、M. 卡雷拉、M. –C. 里亚齐、P. 贝弗林、J. 奥迪（"建了一座桥"）、S. 罗伯茨、B. 弗利弗伯格、E. 博佳德、P. 博格西安、S. 赖利、G. 奥里吉、S. 安蒙斯等（我有时会记起那些对我非常有帮助的人的名字，而在那时表达感激之情为时已晚，在这里先行谢过）。

ACKNOWLEDGMENTS

P. Tanous, L. de Chantal, B. Oppetit, M. Blyth, N. Vardy, B. Appleyard, C. Mihailescu, J. Baz, B. Dupire, Y. Zilber, S. Roberts, A. Pilpel, W. Goodlad, W. Murphy, M. Brockman, J. Brockman, C. Taleb, C. Sandis, J. Kujat, T. Burnham, R. Dobelli, M. Ghosn (the younger), S. Taleb, D. Riviere, J. Gray, M. Carreira, M.-C. Riachi, P. Bevelin, J. Audi (pontem fecit), S. Roberts, B. Flyvberg, E. Boujaoude, P. Boghossian, S. Riley, G. Origgi, S. Ammons, and many more (I sometimes remember names of critically helpful people when it is too late to show gratitude).

who predicted that I would see the light and grow up to make a career in poetry, once I got this ideas business out of my system. More recently, readers have triggered numerous copyright alerts by posting quotes from my books on the Web, but I had never thought of re-expressing my ideas (or, rather, my central idea about the limits of knowledge) in such a way until I realized that these sentences come naturally to me, almost involuntarily, in an eerie way, particularly when walking (slowly) or when freeing up my mind to do nothing, or nothing effortful—I could convince myself that I was hearing voices from the other side of the veil of opacity.

By setting oneself totally free of constraints, free of thoughts, free of this debilitating activity called work, free of efforts, elements hidden in the texture of reality start staring at you; then mysteries that you never thought existed emerge in front of your eyes.

us, convincingly, that he was the greatest Arab poet), for satires[*] (Martial, Aesop, Almaarri), by the moralistes (Vauvenargues, La Rochefoucauld, La Bruyère, Chamfort), to expose opaque philosophy (Wittgenstein), relatively clearer ones (Schopenhauer, Nietzsche, Cioran), or crystal-clear ideas (Pascal).[**] You never have to explain an aphorism—like poetry, this is something that the reader needs to deal with by himself.[***]

There are bland aphorisms, the platitudinous ones harboring important truths that you had thought about before (the kind that make intelligent people recoil at Gibran's *The Prophet*); pleasant ones, those you never thought about but trigger in you the Aha! of an important discovery (such as those in La Rochefoucauld); but the best are those you did not think about before, and for which it takes you more than one reading to realize that they are important truths, particularly when the silent character of the truth in them is so powerful that they are forgotten as soon as read.

Aphorisms require us to change our reading habits and approach them in small doses; each one of them is a complete unit, a complete narrative dissociated from others.

My best definition of a nerd: someone who asks you to explain an aphorism.

I have been aware that my style was aphoristic. As a teenager, I was mentored by the poet Georges Schéhadé (his poetry reads like proverbs),

---

[*] The best way to measure the loss of intellectual sophistication in the Internet age—this "nerdification," to put it bluntly—is in the growing disappearance of sarcasm, as mechanistic minds take insults a bit too literally.

[**] It is not uncommon to find the same maxim repeated by several authors separated by a millennium or a continent.

[***] The aphorism has been somewhat debased (outside the German language) by its association with witticism, such as the ones by Oscar Wilde, Mark Twain, Ambrose Bierce, or Sacha Guitry—deep thought can be poetic and witty, as with Schopenhauer, Nietzsche, or (sometimes) Wittgenstein; but, abiding by the distinction between Sacred and Profane, philosophy and poetry are not stand-up comedy.

market version),* with some show of bravado in the ability of the author to compress powerful ideas in a handful of words—particularly in an oral format. Indeed, it had to be bravado, because the Arabic word for an improvised one-liner is "act of manliness," though such a notion of "manliness" is less gender-driven than it sounds and can be equally translated as "the skills of being human" (virtue has the same roots in Latin, vir, "man"). As if those who could produce powerful thoughts in such a way were invested with talismanic powers.

This mode is at the center of the Levantine soul (and the broader Eastern Mediterranean). When God spoke to the Semites, he spoke in very short poetic sentences, usually through the mouths of prophets. Consider the Scriptures, more particularly the books of Proverbs and Ecclesiastes; Islam's holy book, the Koran, is a collection of concentrated aphorisms. And the format has been adopted for synthetic literary prophecies: Nietzsche's *Zarathustra*, or, more recently, my compatriot from a neighboring (and warring) village in northern Lebanon, Kahlil Gibran, author of *The Prophet*.

Outside of what we now call religion, take the aphorisms of Heraclitus and Hippocrates; the works of Publilius Syrus (a Syrian slave who owed his freedom to his eloquence, expressed in his *Sententiae*, potent one-line poems that echo in the maxims of La Rochefoucauld), and the poetry of the poet who is broadly considered the greatest of all Arab poets, al-Mutanabbi.

Aphorisms as stand-alone sentences have been used for exposition, for religious text, for advice to a grandchild by a Levantine grandmother, for boasting (as I said earlier, in an aphorism, al-Mutanabbi used them to tell

---

* Note the distinction from TV one-liners: the sound bite loses information; the aphorism gains. Somehow, aphorisms obey the Gigerenzer and Goldstein "less is more" effect.

"rational" models, have tried to change humans to fit technology, fudged our ethics to fit our needs for employment, asked economic life to fit the theories of economists, and asked human life to squeeze into some narrative.

We are robust when errors in the representation of the unknown and understanding of random effects do not lead to adverse outcomes—fragile otherwise. The robust benefits from Black Swan events,* the fragile is severely hit by them. We are more and more fragile to a certain brand of scientific autism making confident claims about the unknown—leading to expert problems, risk, massive dependence on human error. As the reader can see from my aphorisms, I have respect for mother nature's methods of robustness (billions of years allow most of what is fragile to break); classical thought is more robust (in its respect for the unknown, the epistemic humility) than the modern post-Enlightenment naïve pseudoscientific autism. Thus my classical values make me advocate the triplet of erudition, elegance, and courage; against modernity's phoniness, nerdiness, and philistinism.**

Art is robust; science, not always (to put it mildly). Some Procrustean beds make life worth living: art and, the most potent of all, the poetic aphorism.

Aphorisms, maxims, proverbs, short sayings, even, to some extent, epigrams are the earliest literary form—often integrated into what we now call poetry. They carry the cognitive compactness of the sound bite (though both more potent and more elegant than today's down-

---

* A Black Swan (capitalized) is an event (historical, economic, technological, personal) that is both unpredicted by some observer and carries massive consequences. In spite of growth in our knowledge, the role of these Black Swans has been growing.

** Many philistines reduce my ideas to an opposition to technology when in fact I am opposing the naïve blindness to its side effects—the fragility criterion. I'd rather be unconditional about ethics and conditional about technology than the reverse.

designed to deal with complexity and nonlinear uncertainties.* Counter to the common discourse, more information means more delusions: our detection of false patterns is growing faster and faster as a side effect of modernity and the information age: there is this mismatch between the messy randomness of the information-rich current world, with its complex interactions, and our intuitions of events, derived in a simpler ancestral habitat. Our mental architecture is at an increased mismatch with the world in which we live.

This leads to sucker problems: when the map does not correspond to the territory, there is a certain category of fool—the overeducated, the academic, the journalist, the newspaper reader, the mechanistic "scientist," the pseudo-empiricist, those endowed with what I call "epistemic arrogance," this wonderful ability to discount what they did not see, the unobserved—who enter a state of denial, imagining the territory as fitting his map. More generally, the fool here is someone who does the wrong reduction for the sake of reduction, or removes something essential, cutting off the legs, or, better, part of the head of a visitor while insisting that he preserved his persona with 95 percent accuracy. Look around at the Procrustean beds we've created, some beneficial, some more questionable: regulations, top-down governments, academia, gyms, commutes, high-rise office buildings, involuntary human relationships, employment,etc.

Since the Enlightenment, in the great tension between rationalism (how we would like things to be so they make sense to us) and empiricism (how things are), we have been blaming the world for not fitting the beds of

---

* Nor is science capable of dealing effectively with nonlinear and complex matters, those fraught with interdependence (climate, economic life, the human body), in spite of its hyped-up successes in the linear domain (physics and engineering), which give it a prestige that has endangered us.

POSTFACE

The general theme of my work is the limitations of human knowledge, and the charming and less charming errors and biases when working with matters that lie outside our field of observation, the unobserved and the unobservables—the unknown; what lies on the other side of the veil of opacity.

Because our minds need to reduce information, we are more likely to try to squeeze a phenomenon into the Procrustean bed of a crisp and known category (amputating the unknown), rather than suspend categorization, and make it tangible. Thanks to our detections of false patterns, along with real ones, what is random will appear less random and more certain—our overactive brains are more likely to impose the wrong, simplistic narrative than no narrative at all.[*]

The mind can be a wonderful tool for self-delusion—it was not

---

[*] This discounting of the unseen comes from the human "scorn of the abstract" (our minds are not good at handling the non-anecdotal and tend to be swayed by vivid imagery, making the media distort our view of the world).

THE END

Platonic minds expect life to be like film, with defined terminal endings; a-Platonic ones expect film to be like life and, except for a few irreversible conditions such as death, distrust the terminal nature of all human-declared endings.

Athenian drinking party in which nonnerds talked about love; alas, there was no drinking and, mercifully, nobody talked about love.

You will get the most attention from those who hate you. No friend, no admirer, and no partner will flatter you with as much curiosity.

When a young woman partners with an otherwise uninteresting rich man, she can sincerely believe that she is attracted to some very specific body part (say, his nose, neck, or knee).

A good foe is far more loyal, far more predictable, and, to the clever, far more useful than the most valuable admirer.

If my detractors knew me better they would hate me even more.

At any stage, humans can thirst for money, knowledge, or love; sometimes for two, never for three.

Love without sacrifice is like theft.

Marriage is the institutional process of feminizing men—and feminizing women.

There are men who surround themselves with women (and seek wealth) for ostentation; others who do so mostly for consumption; they are rarely the same.

Outside of friendship and love, it is very hard to find situations with bilateral, two-way suckers.

I attended a symposium, an event named after a fifth-century (B.C.)

ON THE VARIETIES OF LOVE AND NONLOVE

means attractive.

What organized dating sites fail to understand is that people are far more interesting in what they don't say about themselves.

For company, you often prefer those who find you interesting over those you find interesting.

The Internet broke the private-public wall; impulsive and inelegant utterances that used to be kept private are now available for literal interpretation.

One of the problems with social networks is that it is getting harder and harder for others to complain about you behind your back.

You can be certain that a person has the means but not the will to help you when he says "there is nothing else I can do." And you can be certain that a person has neither means nor will to help you when he says "I am here to help."

We expect places and products to be less attractive than in marketing brochures, but we never forgive humans for being worse than their first impressions.

When someone starts a sentence with "simply," you should expect to hear something very complicated.

Half the people lie with their lips; the other half with their tears.

You know you have influence when people start noticing your absence more than the presence of others.

You are guaranteed a repetition when you hear the declaration "never again!"

Some reticent people use silence to conceal their intelligence; but most do so to hide the lack of it.

When someone says "I am not that stupid," it often means that he is more stupid than he thinks.

Bad-mouthing is the only genuine, never faked expression of admiration.

When a woman says about a man that he is intelligent, she often means handsome; when a man says about a woman that she is dumb, he always

THE IMPLICIT AND THE EXPLICIT

The two most celebrated acts of courage in history aren't Homeric fighters but two Eastern Mediterranean fellows who died, even sought death, for their ideas.

The weak cannot be good; or, perhaps, he can only be good within an exhaustive and overreaching legal system.

By all means, avoid words—threats, complaints, justification, narratives, reframing, attempts to win arguments, supplications; avoid words!

According to Lucian of Samosata, the philosopher Demonax stopped a Spartan from beating his servant. "You are making him your equal," he said.

The classical man's worst fear was inglorious death; the modern man's worst fear is just death.

without being wise.*

The traits I respect are erudition and the courage to stand up when half-men are afraid for their reputation. Any idiot can be intelligent.

The mediocre regret their words more than their silence; finer men regret their silence more than their words; the magnificent has nothing to regret.

Regular men are a certain varying number of meals away from lying, stealing, killing, or even working as forecasters for the Federal Reserve in Washington; never the magnificent.**

Social science means inventing a certain brand of human we can understand.

When expressing "good luck" to a peer, the weak wishes the opposite; the strong is mildly indifferent; but only the magnificent means it.

In the past, only some of the males, but all of the females, were able to procreate. Equality is more natural for females.

The magnificent believes half of what he hears and twice what he says.

A verbal threat is the most authentic certificate of impotence.

---

* Looking at Federal Reserve Chairman Ben Bernanke.

** I had to read Aristotle's *Nicomachean Ethics* Book IV ten times before realizing what he didn't say explicitly (but knew): the magnificent (megalopsychos) is all about unconditionals.

Mediocre men tend to be outraged by small insults but passive, subdued, and silent in front of very large ones.*

The only definition of an alpha male: if you try to be an alpha male, you will never be one.

Those who have nothing to prove never say that they have nothing to prove.

The weak shows his strength and hides his weaknesses; the magnificent exhibits his weaknesses like ornaments.

How superb to become wise without being boring; how sad to be boring

---

* Consider the reaction to the banking and economics establishments.

THE SAGE, THE WEAK, AND THE MAGNIFICENT*

---

* In Aristotle's *Nicomachean Ethics*, the megalopsychos, which I translate as the magnificent, is the "great-souled" who thinks of himself as worthy of great things and, aware of his own position in life, abides by a certain system of ethics that excludes pettiness. This notion of great soul, though displaced by Christian ethics advocating humility, remains present in Levantine culture, with the literal *Kabir al-nafs*. Among other attributes, the magnificent walks slowly.

We should make students recompute their GPAs by counting their grades in finance and economics backward.

The agency problem drives every company, thanks to the buildup of hidden risks, to maximal fragility.

In politics we face the choice between warmongering, nation-state-loving, big-business agents on one hand; and risk-blind, top-down, epistemic arrogant big servants of large employers on the other. But we have a choice.

"It is much easier to scam people for billions than for just millions."[*]

At a panel in Moscow, I watched the economist Edmund Phelps, who got the "Nobel" for writings no one reads, theories no one uses, and lectures no one understands.

One of the failures of "scientific approximation" in the nonlinear domain comes from the inconvenient fact that the average of expectations is different from the expectation of averages. Don't cross a river, because it is on average four feet deep. This is also known as Jensen's inequality.

Journalists as reverse aphorists: my statement "you need skills to get a BMW, skills plus luck to become a Warren Buffett" was summarized as "Taleb says Buffett has no skills."

The curious mind embraces science; the gifted and sensitive, the arts; the practical, business; the leftover becomes an economist.

Public companies, like human cells, are programmed for apoptosis, suicide through debt and hidden risks. Bailouts invest the process with a historical dimension.

In poor countries, officials receive explicit bribes; in D.C. they get the sophisticated implicit, unspoken promise to work for large corporations.

Fate is at its cruelest when a banker ends up in poverty.

---

[*]   Inspired by the Madoff episode.

the right believes that because models are stupid markets should be smart. Alas, it never hit both sides that both markets and models are very stupid.

Economics is like a dead star that still seems to produce light; but you know it is dead.

Suckers think that you cure greed with money, addiction with substances, expert problems with experts, banking with bankers, economics with economists, and debt crises with debt spending.

You can be certain that the head of a corporation has a lot to worry about when he announces publicly that "there is nothing to worry about."

The stock market, in brief: participants are calmly waiting in line to be slaughtered while thinking it is for a Broadway show.

The main difference between government bailouts and smoking is that in some rare cases the statement "this is my last cigarette" holds true.

What makes us fragile is that institutions cannot have the same virtues (honor, truthfulness, courage, loyalty, tenacity) as individuals.

The worst damage has been caused by competent people trying to do good; the best improvements have been brought by incompetent ones not trying to do good.

The difference between banks and the Mafia: banks have better legal-regulatory expertise, but the Mafia understands public opinion.

There are designations, like "economist," "prostitute," or "consultant," for which additional characterization doesn't add information.

A mathematician starts with a problem and creates a solution; a consultant starts by offering a "solution" and creates a problem.

What they call "risk" I call opportunity; but what they call "low risk" opportunity I call sucker problem.

Organizations are like caffeinated dupes unknowingly jogging backward; you only hear of the few who reach their destination.

The best test of whether someone is extremely stupid (or extremely wise) is whether financial and political news makes sense to him.

The left holds that because markets are stupid models should be smart;

ECONOMIC LIFE AND OTHER VERY VULGAR SUBJECTS

exactly the salaried researcher and the modern tenure-loving academic. Progress.

Engineers can compute but not define, mathematicians can define but not compute, economists can neither define nor compute.

Something finite but with unknown upper bounds is epistemically equivalent to something infinite. This is epistemic infinity.

Conscious ignorance, if you can practice it, expands your world; it can make things infinite.

For the classics, philosophical insight was the product of a life of leisure; for me, a life of leisure is the product of philosophical insight.

It takes a lot of intellect and confidence to accept that what makes sense doesn't really make sense.

A theological Procrustean bed: for the Orthodox since Gregory Palamas and for the Arabs since Algazel, attempts to define God using the language of philosophical universals were a rationalistic mistake. I am still waiting for a modern to take notice.

Saying "the mathematics of uncertainty" is like saying "the chastity of sex"—what is mathematized is no longer uncertain, and vice versa.

Sadly, we learn the most from fools, economists, and other reverse role models, yet we pay them back with the worst ingratitude.

In Plato's *Protagoras*, Socrates contrasts philosophy as the collaborative search for truth with the sophist's use of rhetoric to gain the upper hand in argument for fame and money. Twenty-five centuries later, this is

To become a philosopher, start by walking very slowly.

Real mathematicians understand completeness, real philosophers understand incompleteness, the rest don't formally understand anything.

In twenty-five centuries, no human came along with the brilliance, depth, elegance, wit, and imagination to match Plato—to protect us from his legacy.

Why do I have an obsessive Plato problem? Most people need to surpass their predecessors; Plato managed to surpass all his successors.

To be a philosopher is to know through long walks, by reasoning, and reasoning only, a priori, what others can only potentially learn from their mistakes, crises, accidents, and bankruptcies—that is, a posteriori.

BEING A PHILOSOPHER AND MANAGING TO REMAIN ONE

For Seneca, the Stoic sage should withdraw from public efforts when unheeded and the state is corrupt beyond repair. It is wiser to wait for self-destruction.

A prophet is not someone with special visions, just someone blind to most of what others see.

For the ancients, forecasting historical events was an insult to the God(s); for me, it is an insult to man—that is, for some, to science.

The ancients knew very well that the only way to understand events was to cause them.

Anyone voicing a forecast or expressing an opinion without something at risk has some element of phoniness. Unless he risks going down with the ship this would be like watching an adventure movie.

They would take forecasting more seriously if it were pointed out to them that in Semitic languages the words for forecast and "prophecy" are the same.

THE SCANDAL

OF

PREDICTION

what does not work, what not to do), not what we add (what to do).*

They think that intelligence is about noticing things that are relevant (detecting patterns); in a complex world, intelligence consists in ignoring things that are irrelevant (avoiding false patterns).

Happiness; we don't know what it means, how to measure it, or how to reach it, but we know extremely well how to avoid unhappiness.

The imagination of the genius vastly surpasses his intellect; the intellect of the academic vastly surpasses his imagination.

The ideal trivium education, and the least harmful one to society and pupils, would be mathematics, logic, and Latin; a double dose of Latin authors to compensate for the severe loss of wisdom that comes from mathematics; just enough mathematics and logic to control verbiage and rhetoric.

The four most influential moderns: Darwin, Marx, Freud, and (the productive) Einstein were scholars but not academics. It has always been hard to do genuine—and nonperishable—work within institutions.

---

* The best way to spot a charlatan: someone (like a consultant or a stockbroker) who tells you what to do instead of what not to do.

Since Plato, Western thought and the theory of knowledge have focused on the notions of True-False; as commendable as it was, it is high time to shift the concern to Robust-Fragile, and social epistemology to the more serious problem of Sucker-Nonsucker.

The problem of knowledge is that there are many more books on birds written by ornithologists than books on birds written by birds and books on ornithologists written by birds.

The perfect sucker understands that pigs can stare at pearls but doesn't realize he can be in an analog situation.

It takes extraordinary wisdom and self-control to accept that many things have a logic we do not understand that is smarter than our own.

Knowledge is subtractive, not additive—what we subtract (reduction by

EPISTEMOLOGY AND SUBTRACTIVE KNOWLEDGE

tell who is heroic and who is not.

I suspect that IQ, SAT, and school grades are tests designed by nerds so they can get high scores in order to call each other intelligent.*

They read Gibbon's *Decline and Fall* on an eReader but refuse to drink Château Lynch-Bages in a Styrofoam cup.

My best example of the domain dependence of our minds, from my recent visit to Paris: at lunch in a French restaurant, my friends ate the salmon and threw away the skin; at dinner, at a sushi bar, the very same friends ate the skin and threw away the salmon.

Fragility: we have been progressively separating human courage from warfare, allowing wimps with computer skills to kill people without the slightest risk to their lives.

---

* Smart and wise people who score low on IQ tests, or patently intellectually defective ones, like former U.S. president George W. Bush, who score high on them (130), are testing the test and not the reverse.

Sports are commoditized and, alas, prostituted randomness.

When you beat up someone physically, you get exercise and stress relief; when you assault him verbally on the Internet, you just harm yourself.

Just as smooth surfaces, competitive sports, and specialized work fossilize mind and body, competitive academia fossilizes the soul.

They agree that chess training only improves chess skills but disagree that classroom training (almost) only improves classroom skills.

Upon arriving at the hotel in Dubai, the businessman had a porter carry his luggage; I later saw him lifting free weights in the gym.

Games were created to give nonheroes the illusion of winning. In real life, you don't know who really won or lost (except too late), but you can

THE LUDIC FALLACY AND DOMAIN DEPENDENCE*

---

* Ludic is Latin for "related to games"; the fallacy prevalent in *The Black Swan* about making life resemble games (or formal setups) with crisp rules rather than the reverse. Domain dependence is when one acts in a certain way in an environment (say, the gym) and a different way in another.

deficits tend to be larger, rarely smaller, than planned.

Robust is when you care more about the few who like your work than the multitude who dislike it (artists); fragile when you care more about the few who dislike your work than the multitude who like it (politicians).

The rationalist imagines an imbecile-free society; the empiricist an imbecile-proof one, or, even better, a rationalist-proof one.

Academics are only useful when they try to be useless (say, as in mathematics and philosophy) and dangerous when they try to be useful.

For the robust, an error is information; for the fragile, an error is an error.

The best test of robustness to reputational damage is your emotional state (fear, joy, boredom) when you get an email from a journalist.

The main disadvantage of being a writer, particularly in Britain, is that there is nothing you can do in public or private that would damage your reputation.

Passionate hate (by nations and individuals) ends by rotation to another subject of hate; mediocrity cannot handle more than one enemy. This makes warring statelings with shifting alliances and enmities a robust system.

I find it inconsistent (and corrupt) to dislike big government while favoring big business—but (alas) not the reverse.

How often have you arrived one, three, or six hours late on a transatlantic flight as opposed to one, three, or six hours early? This explains why

You are only secure if you can lose your fortune without the additional worse insult of having to become humble.*

To test someone's robustness to reputational errors, ask a man in front of an audience if he is "still doing poorly" or if he is "still losing money" and watch his reaction.

Robustness is progress without impatience.

When conflicted between two choices, take neither.

Nation-states like war; city-states like commerce; families like stability; and individuals like entertainment.

---

* My great-great-great-great-great grandfather's rule.

ROBUSTNESS AND FRAGILITY

In Proust there is a character, Morel, who demonizes Nissim Bernard, a Jew who lent him money, and becomes anti-Semitic just so he can escape the feeling of gratitude.

Promising someone good luck as a reward for good deeds sounds like a bribe—perhaps the remnant of an archaic, pre-deontic pre-classical morality.

The difference between magnificence and arrogance is in what one does when nobody is looking.

The nation-state: apartheid without political incorrectness.

In a crowd of a hundred, 50 percent of the wealth, 90 percent of the imagination, and 100 percent of the intellectual courage will reside in a single person—not necessarily the same one.

Just as dyed hair makes older men less attractive, it is what you do to hide your weaknesses that makes them repugnant.

For soldiers, we use the term "mercenary," but we absolve employees of responsibility with "everybody needs to make a living."

English does not distinguish between arrogant-up (irreverence toward the temporarily powerful) and arrogant-down (directed at the small guy).

Someone from your social class who becomes poor affects you more than thousands of starving ones outside of it.

You may outlive your strength, never your wisdom.

Weak men act to satisfy their needs, stronger men their duties.

Religions and ethics have evolved from promising heaven if you do good, to promising heaven while you do good, to making you promise to do good.

Avoid calling heroes those who had no other choice.

There are those who will thank you for what you gave them and others who will blame you for what you did not give them.

Ethical man accords his profession to his beliefs, instead of according his beliefs to his profession. This has been rarer and rarer since the Middle Ages.

I trust everyone except those who tell me they are trustworthy.

People often need to suspend their self-promotion, and have someone in their lives they do not need to impress. This explains dog ownership.

Pure generosity is when you help the ingrate. Every other form is self-serving.*

I wonder if crooks can conceive that honest people can be shrewder than they.

---

* Kantian ethics.

she was at the first encounter and the most recent one.

Meditation is a way to be narcissistic without hurting anyone.

True humility is when you can surprise yourself more than others; the rest is either shyness or good marketing.

We find it to be in extremely bad taste for individuals to boast of their accomplishments; but when countries do so we call it "national pride."

You can only convince people who think they can benefit from being convinced.

Greatness starts with the replacement of hatred with polite disdain.

Trust people who make a living lying down or standing up more than those who do so sitting down.

The tragedy of virtue is that the more obvious, boring, unoriginal, and sermonizing the proverb, the harder it is to implement.

Even the cheapest misers can be generous with advice.

If you lie to me, keep lying; don't hurt me by suddenly telling the truth.

Don't trust a man who needs an income—except if it is minimum wage.[*]

---

[*] Those in corporate captivity would do anything to "feed a family."

If you find any reason why you and someone are friends, you are not friends.

My biggest problem with modernity may lie in the growing separation of the ethical and the legal.*

Life's beauty: the kindest act toward you in your life may come from an outsider not interested in reciprocation.**

We are most motivated to help those who need us the least.

To value a person, consider the difference between how impressive he or

---

* Former U.S. Treasury secretary "bankster" Robert Rubin, perhaps the biggest thief in history, broke no law. The difference between legal and ethical increases in a complex system ... then blows it up.

** The flip side: the worst pain inflicted on you will come from someone who at some point in your life cared about you.

ETHICS

Al-Mutanabbi boasted that he was the greatest of all Arab poets, but he said so in the greatest of all Arab poems.

Wit seduces by signaling intelligence without nerdiness.

In classical renderings of prominent figures, males are lean and females are plump; in modern photographs, the opposite.

Just as no monkey is as good-looking as the ugliest of humans, no academic is worthier than the worst of the creators.

If you want to annoy a poet, explain his poetry.

Art is a one-sided conversation with the unobserved.

The genius of Benoît Mandelbrot is in achieving aesthetic simplicity without having recourse to smoothness.

Beauty is enhanced by unashamed irregularities; magnificence by a façade of blunder.

To understand "progress": all places we call ugly are both man-made and modern (Newark), never natural or historical (Rome).

We love imperfection, the right kind of imperfection; we pay up for original art and typo-laden first editions.

Most people need to wait for another person to say "this is beautiful art" to say "this is beautiful art"; some need to wait for two or more.

AESTHETICS

ones.

Randomness is indistinguishable from complicated, undetected, and undetectable order; but order itself is indistinguishable from artful randomness.

views himself as more generic and others more unique.

What made medicine fool people for so long was that its successes were prominently displayed and its mistakes (literally) buried.

The sucker's trap is when you focus on what you know and what others don't know, rather than the reverse.

Medieval man was a cog in a wheel he did not understand; modern man is a cog in a complicated system he thinks he understands.

The calamity of the information age is that the toxicity of data increases much faster than its benefits.

The role of the media is best seen in the journey from Cato the Elder to a modern politician.* Do some extrapolation if you want to be scared.

Mental clarity is the child of courage, not the other way around.**

Most info-Web-media-newspaper types have a hard time swallowing the idea that knowledge is reached (mostly) by removing junk from people's heads.

Finer men tolerate others' small inconsistencies though not the large ones; the weak tolerate others' large inconsistencies though not small

---

* Say, Sarah Palin.

** The biggest error since Socrates has been to believe that lack of clarity is the source of all our ills, not the result of them.

Unless we manipulate our surroundings, we have as little control over what and whom we think about as we do over the muscles of our hearts.

Corollary to Moore's Law: every ten years, collective wisdom degrades by half.[*]

Never rid anyone of an illusion unless you can replace it in his mind with another illusion. (But don't work too hard on it; the replacement illusion does not even have to be more convincing than the initial one.)

The tragedy is that much of what you think is random is in your control and, what's worse, the opposite.

The fool views himself as more unique and others more generic; the wise

---

[*] Moore's Law stipulates that computational power doubles every eighteen months.

FOOLED BY RANDOMNESS

harm you in some circumstances. The more complex the system, the weaker the notion of Universal.

The fool generalizes the particular; the nerd particularizes the general; some do both; and the wise does neither.

You want to be yourself, idiosyncratic; the collective (school, rules, jobs, technology) wants you generic to the point of castration.

True love is the complete victory of the particular over the general, and the unconditional over the conditional.

What I learned on my own I still remember.

Regular minds find similarities in stories (and situations); finer minds detect differences.

To grasp the difference between Universal and Particular, consider that some dress better to impress a single, specific person than an entire crowd.

We unwittingly amplify commonalities with friends, dissimilarities with strangers, and contrasts with enemies.

Many are so unoriginal they study history to find mistakes to repeat.

There is nothing deemed harmful (in general) that cannot be beneficial in some particular instances, and nothing deemed beneficial that cannot

THE UNIVERSAL AND THE PARTICULAR

The information-rich Dark Ages: in 2010, 600,000 books were published, just in English, with few memorable quotes. Circa AD zero, a handful of books were written. In spite of the few that survived, there are loads of quotes.

In the past, most were ignorant, one in a thousand were refined enough to talk to. Today, literacy is higher, but thanks to progress, the media, and finance, only one in ten thousand.

We are better at (involuntarily) doing out of the box than (voluntarily) thinking out of the box.

Half of suckerhood is not realizing that what you don't like might be loved by someone else (hence by you, later), and the reverse.

It is much less dangerous to think like a man of action than to act like a man of thought.

Literature comes alive when covering up vices, defects, weaknesses, and confusions; it dies with every trace of preaching.

I can predict when an author is about to plagiarize me, and poorly so when he writes that Taleb "popularized" the theory of Black Swan events.*

Newspaper readers exposed to real prose are like deaf persons at a Puccini opera: they may like a thing or two while wondering, "what's the point?"

Some books cannot be summarized (real literature, poetry); some can be compressed to about ten pages; the majority to zero pages.

The exponential information age is like a verbally incontinent person: he talks more and more as fewer and fewer people listen.

What we call fiction is, when you look deep, much less fictional than nonfiction; but it is usually less imaginative.

It's much harder to write a book review for a book you've read than for a book you haven't read.

Most so-called writers keep writing and writing with the hope to, some day, find something to say.

Today, we mostly face the choice between those who write clearly about a subject they don't understand and those who write poorly about a subject they don't understand.

---

* It is also an indicator that he will imitate, "me, too" style, my business.

written by academics, read the footnotes and skip the text; and with business books, skip both the text and the footnotes.

Double a man's erudition; you will halve his citations.

Losers, when commenting on the works of someone patently more impressive, feel obligated to unnecessarily bring down their subject by expressing what he is not ("he is not a genius, but ..."; "while he is no Leonardo ...") instead of expressing what he is.

You are alive in inverse proportion to the density of clichés in your writing.

What we call "business books" is an eliminative category invented by bookstores for writings that have no depth, no style, no empirical rigor, and no linguistic sophistication.

Just like poets and artists, bureaucrats are born, not made; it takes normal humans extraordinary effort to keep attention on such boring tasks.

The costs of specialization: architects build to impress other architects; models are thin to impress other models; academics write to impress other academics; filmmakers try to impress other filmmakers; painters impress art dealers; but authors who write to impress book editors tend to fail.

It is a waste of emotions to answer critics; better to stay in print long after they are dead.

defects.

For pleasure, read one chapter by Nabokov. For punishment, two.

There is a distinction between expressive hypochondria and literature, just as there is one between self-help and philosophy.

You need to keep reminding yourself of the obvious: charm lies in the unsaid, the unwritten, and the undisplayed. It takes mastery to control silence.

No author should be considered as having failed until he starts teaching others about writing.

Hard science gives sensational results with a horribly boring process; philosophy gives boring results with a sensational process; literature gives sensational results with a sensational process; and economics gives boring results with a boring process.

A good maxim allows you to have the last word without even starting a conversation.

Just as there are authors who enjoy having written and others who enjoy writing, there are books you enjoy reading and others you enjoy having read.

A genius is someone with flaws harder to imitate than his qualities.

With regular books, read the text and skip the footnotes; with those

Writing is the art of repeating oneself without anyone noticing.

Most people write so they can remember things; I write to forget.

What they call philosophy I call literature; what they call literature I call journalism; what they call journalism I call gossip; and what they call gossip I call (generously) voyeurism.

Writers are remembered for their best work, politicians for their worst mistakes, and businessmen are almost never remembered.

Critics may appear to blame the author for not writing the book they wanted to read; but in truth they are blaming him for writing the book they wanted, but were unable, to write.

Literature is not about promoting qualities, rather, airbrushing (your)

THE REPUBLIC

OF

LETTERS

When I look at people on treadmills I wonder how alpha lions, the strongest, expend the least amount of energy, sleeping twenty hours a day; others hunt for them. Caesar pontem fecit.*

Every social association that is not face-to-face is injurious to your health.

---

* Literally, "Caesar built a bridge," but the subtlety is that it can also suggest that "he had a bridge built for him."

For everything, use boredom in place of a clock, as a biological wristwatch, though under constraints of politeness.

Decomposition, for most, starts when they leave the free, social, and uncorrupted college life for the solitary confinement of professions and nuclear families.

For a classicist, a competitive athlete is painful to look at; trying hard to become an animal rather than a man, he will never be as fast as a cheetah or as strong as an ox.

Skills that transfer: street fights, off-path hiking, seduction, broad erudition. Skills that don't: school, games, sports, laboratory—what's reduced and organized.

You exist in full if and only if your conversation (or writings) cannot be easily reconstructed with clips from other conversations.

The English have random Mediterranean weather; but they go to Spain because their free hours aren't free.

For most, work and what comes with it have the eroding effect of chronic injury.

Technology is at its best when it is invisible.

The difference between true life and modern life equals the one between a conversation and bilateral recitations.

while convincing him that it is becoming more "efficient."

The difference between technology and slavery is that slaves are fully aware that they are not free.

You have a real life if and only if you do not compete with anyone in any of your pursuits.

With terminal disease, nature lets you die with abbreviated suffering; medicine lets you suffer with prolonged dying.

We are satisfied with natural (or old) objects like vistas or classical paintings but insatiable with technologies, amplifying small improvements in versions, obsessed about 2.0, caught in a mental treadmill.

Only in recent history has "working hard" signaled pride rather than shame for lack of talent, finesse, and, mostly, sprezzatura.

Their idea of the sabbatical is to work six days and rest for one; my idea of the sabbatical is to work for (part of) a day and rest for six.

What they call "play" (gym, travel, sports) looks like work; the harder they try, the more captive they are.

Most modern efficiencies are deferred punishment.

We are hunters; we are only truly alive in those moments when we improvise; no schedule, just small surprises and stimuli from the environment.

The three most harmful addictions are heroin, carbohydrates, and a monthly salary.

My only measure of success is how much time you have to kill.

I wonder if a lion (or a cannibal) would pay a high premium for free-range humans.

If you need to listen to music while walking, don't walk; and please don't listen to music.

Men destroy each other during war; themselves during peacetime.

Sports feminize men and masculinize women.

Technology can degrade (and endanger) every aspect of a sucker's life

THESEUS, OR LIVING THE PALEO LIFE

I need to keep reminding myself that a truly independent thinker may look like an accountant.

Mathematics is to knowledge what an artificial hand is to the real one; some amputate to replace.

Modernity inflicts a sucker narrative on activities; now we "walk for exercise," not "walk" with no justification; for hidden reasons.

Social media are severely antisocial, health foods are empirically unhealthy, knowledge workers are very ignorant, and social sciences aren't scientific at all.

For so many, instead of looking for "cause of death" when they expire, we should be looking for "cause of life" when they are still around.

It is those who use others who are the most upset when someone uses them.

If someone gives you more than one reason why he wants the job, don't hire him.

Failure of second-order thinking: he tells you a secret and somehow expects you to keep it, when he just gave you evidence that he can't keep it himself.

Social networks present information about what people like; more informative if, instead, they described what they don't like.

People are so prone to overcausation that you can make the reticent turn loquacious by dropping an occasional "why?" in the conversation.

The most depressing aspect of the lives of the couples you watch surreptitiously arguing in restaurants is that they are almost always unaware of the true subject of argument.

It seems that it is the most unsuccessful people who give the most advice, particularly for writing and financial matters.

Rumors are only valuable when they are denied.

Over the long term, you are more likely to fool yourself than others.

There are two types of people: those who try to win and those who try to win arguments. They are never the same.

People usually apologize so they can do it again.

CHARMING AND LESS CHARMING SUCKER PROBLEMS

The Web's "connectedness" creates a peculiar form of informational and pseudosocial promiscuity, which makes one feel clean after Web rationing.

In most debates, people seem to be trying to convince one another; but all they can hope for is new arguments to convince themselves.

I have the fondest memories of time spent in places called ugly, the most boring ones of places called scenic.

Fitness is certainly the sign of strength, but outside of natural stimuli the drive to acquire fitness can signal some deep incurable weakness.

Charm is the ability to insult people without offending them; nerdiness the reverse.

Those who do not think that employment is systemic slavery are either blind or employed.

They are born, then put in a box; they go home to live in a box; they study by ticking boxes; they go to what is called "work" in a box, where they sit in their cubicle box; they drive to the grocery store in a box to buy food in a box; they go to the gym in a box to sit in a box; they talk about thinking "outside the box"; and when they die they are put in a box. All boxes, Euclidian, geometrically smooth boxes.

Another definition of modernity: conversations can be more and more completely reconstructed with clips from other conversations taking place at the same time on the planet.

The twentieth century was the bankruptcy of the social utopia; the twenty-first will be that of the technological one.

Efforts at building social, political, and medical utopias have caused nightmares; many cures and techniques came from martial efforts.

Modernity: we created youth without heroism, age without wisdom, and life without grandeur.

You can tell how uninteresting a person is by asking him whom he finds interesting.

The Web is an unhealthy place for someone hungry for attention.

I wonder if anyone ever measured the time it takes, at a party, before a mildly successful stranger who went to Harvard makes others aware of it.

People focus on role models; it is more effective to find antimodels—people you don't want to resemble when you grow up.

It is a good practice to always apologize, except when you have done something wrong.

Preoccupation with efficacy is the main obstacle to a poetic, noble, elegant, robust, and heroic life.

Some, like most bankers, are so unfit for success that they look like dwarves dressed in giants' clothes.

Don't complain too loud about wrongs done you; you may give ideas to your less imaginative enemies.

Most feed their obsessions by trying to get rid of them.

It is as difficult to change someone's opinions as it is to change his tastes.

Catholic countries had more serial monogamy than today, but without the need for divorce—life expectancy was short; marriage duration was much, much shorter.

The fastest way to become rich is to socialize with the poor; the fastest way to become poor is to socialize with the rich.

You will be civilized on the day you can spend a long period doing nothing, learning nothing, and improving nothing, without feeling the slightest amount of guilt.

Someone who says "I am busy" is either declaring incompetence (and lack of control of his life) or trying to get rid of you.

The difference between slaves in Roman and Ottoman days and today's employees is that slaves did not need to flatter their boss.

You are rich if and only if money you refuse tastes better than money you accept.

For most, success is the harmful passage from the camp of the hating to the camp of the hated.

To see if you like where you are, without the chains of dependence, check if you are as happy returning as you were leaving.

The difference between love and happiness is that those who talk about love tend to be in love, but those who talk about happiness tend to be not happy.

Quite revealing of human preferences that more suicides come from shame or loss of financial and social status than medical diagnoses.

"Wealthy" is meaningless and has no robust absolute measure; use intead the subtractive measure "unwealth," that is, the difference, at any point in time, between what you have and what you would like to have.

Older people are most beautiful when they have what is lacking in the young: poise, erudition, wisdom, phronesis, and this post-heroic absence of agitation.

I went to a happiness conference; researchers looked very unhappy.

What fools call "wasting time" is most often the best investment.

Decline starts with the replacement of dreams with memories and ends with the replacement of memories with other memories.

You want to avoid being disliked without being envied or admired.

Read nothing from the past one hundred years; eat no fruits from the past one thousand years; drink nothing from the past four thousand years (just wine and water); but talk to no ordinary man over forty. A man without a heroic bent starts dying at the age of thirty.

Some pursuits are much duller from the inside. Even piracy, they say.

Karl Marx, a visionary, figured out that you can control a slave much better by convincing him he is an employee.

Success is becoming in middle adulthood what you dreamed to be in late childhood. The rest comes from loss of control.

The opposite of success isn't failure; it is name-dropping.

Modernity needs to understand that being rich and becoming rich are not mathematically, personally, socially, and ethically the same thing.

You don't become completely free by just avoiding to be a slave; you also need to avoid becoming a master.[*]

Fortune punishes the greedy by making him poor and the very greedy by making him rich.

---

[*] Versions of this point have been repeated and rediscovered throughout history—the last convincing one by Montaigne.

CHANCE, SUCCESS, HAPPINESS, AND STOICISM

The book is the only medium left that hasn't been corrupted by the profane: everything else on your eyelids manipulates you with an ad.*

You can replace lies with truth; but myth is only displaced with a narrative.

The sacred is all about unconditionals; the profane is all about conditionals.**

The source of the tragic in history is in mistaking someone else's unconditional for conditional—and the reverse.

Restaurants get you in with food to sell you liquor; religions get you in with belief to sell you rules (e.g., avoid debt). People can understand the notion of God, not unexplained rules, interdicts, and categorical heuristics.

One categorical: it is easier to fast than diet. You cannot be "slightly" kosher or halal by only eating a small portion of ham.

To be completely cured of newspapers, spend a year reading the previous week's newspapers.

---

* A comment here. After a long diet from the media, I came to realize that there is nothing that's not (clumsily) trying to sell you something. I only trust my library. There is nothing wrong with the ownership of the physical book as a manifestation of human weakness, desire to show off, peacock tail-style signaling of superiority, it's the commercial agenda outside the book that corrupts.

** For instance, many people said to be unbribable are just too expensive.

You cannot express the holy in terms made for the profane, but you can discuss the profane in terms made for the holy.

If you can't spontaneously detect (without analyzing) the difference between sacred and profane, you'll never know what religion means. You will also never figure out what we commonly call art. You will never understand anything.

People used to wear ordinary clothes weekdays and formal attire on Sunday. Today it is the exact reverse.

To mark a separation between holy and profane, I take a ritual bath after any contact, or correspondence (even emails), with consultants, economists, Harvard Business School professors, journalists, and those in similarly depraved pursuits; I then feel and act purified from the profane until the next episode.

THE SACRED
AND
THE PROFANE

It is a very recent disease to mistake the unobserved for the nonexistent; but some are plagued with the worse disease of mistaking the unobserved for the unobservable.

Asking science to explain life and vital matters is equivalent to asking a grammarian to explain poetry.

You exist if and only if you are free to do things without a visible objective, with no justification and, above all, outside the dictatorship of someone else's narrative.

MATTERS ONTOLOGICAL

By praising someone for his lack of defects you are also implying his lack of virtues.

When she shouts that what you did was unforgivable, she has already started to forgive you.

Being unimaginative is only a problem when you are easily bored.

We call narcissistic those individuals who behave as if they were the central residents of the world; those who do exactly the same in a set of two we call lovers or, better, "blessed by love."

Friendship that ends was never one; there was at least one sucker in it.

Most people fear being without audiovisual stimulation because they are too repetitive when they think and imagine things on their own.

Unrequited hate is vastly more diminishing for the self than unrequited love. You can't react by reciprocating.

For the compassionate, sorrow is more easily displaced by another sorrow than by joy.

Wisdom in the young is as unattractive as frivolity in the elderly.

Some people are only funny when they try to be serious.

It is difficult to stop the impulse to reveal secrets in conversation, as if information had the desire to live and the power to multiply.

variety of sentiment now called self-esteem.

We ask "why is he rich (or poor)?" not "why isn't he richer (or poorer)?"; "why is the crisis so deep?" not "why isn't it deeper?"

Hatred is much harder to fake than love. You hear of fake love; never of fake hate.

The opposite of manliness isn't cowardice; it's technology.

Usually, what we call a "good listener" is someone with skillfully polished indifference.

It is the appearance of inconsistency, and not its absence, that makes people attractive.

You remember emails you sent that were not answered better than emails that you did not answer.

People reserve standard compliments for those who do not threaten their pride; the others they often praise by calling "arrogant."

Since Cato the Elder, a certain type of maturity has shown up when one starts blaming the new generation for "shallowness" and praising the previous one for its "values."

It is as difficult to avoid bugging others with advice on how to exercise and other health matters as it is to stick to an exercise schedule.

for your looks, for your status—but rarely for your wisdom.

Most of what they call humility is successfully disguised arrogance.

If you want people to read a book, tell them it is overrated.

You never win an argument until they attack your person.

Nothing is more permanent than "temporary" arrangements, deficits, truces, and relationships; and nothing is more temporary than "permanent" ones.

The most painful moments are not those we spend with uninteresting people; rather, they are those spent with uninteresting people trying hard to be interesting.

Hatred is love with a typo somewhere in the computer code, correctable but very hard to find.

I wonder whether a bitter enemy would be jealous if he discovered that I hated someone else.

The characteristic feature of the loser is to bemoan, in general terms, mankind's flaws, biases, contradictions, and irrationality—without exploiting them for fun and profit.

The test of whether you really liked a book is if you reread it (and how many times); the test of whether you really liked someone's company is if you are ready to meet him again and again—the rest is spin, or that

The best revenge on a liar is to convince him that you believe what he said.

When we want to do something while unconsciously certain to fail, we seek advice so we can blame someone else for the failure.

It is harder to say no when you really mean it than when you don't.

Never say no twice if you mean it.

Your reputation is harmed the most by what you say to defend it.

The only objective definition of aging is when a person starts to talk about aging.

They will envy you for your success, for your wealth, for your intelligence,

COUNTER NARRATIVES

If you know, in the morning, what your day looks like with any precision, you are a little bit dead—the more precision, the more dead you are.

There is no intermediate state between ice and water but there is one between life and death: employment.

You have a calibrated life when most of what you fear has the titillating prospect of adventure.

Procrastination is the soul rebelling against entrapment.

Nobody wants to be perfectly transparent; not to others, certainly not to himself.

I wonder if those who advocate generosity for its rewards notice the inconsistency, or if what they call generosity is an attractive investment strategy.*

Those who think religion is about "belief" don't understand religion, and don't understand belief.

Work destroys your soul by stealthily invading your brain during the hours not officially spent working; be selective about professions.

In nature we never repeat the same motion; in captivity (office, gym, commute, sports), life is just repetitive-stress injury. No randomness.

Using, as an excuse, others' failure of common sense is in itself a failure of common sense.

Compliance with the straitjacket of narrow (Aristotelian) logic and avoidance of fatal inconsistencies are not the same thing.

Economics cannot digest the idea that the collective (and the aggregate) are disproportionately less predictable than individuals.

Don't talk about "progress" in terms of longevity, safety, or comfort before comparing zoo animals to those in the wilderness.

---

* A generous act is precisely what should aim at no reward, neither financial nor social nor emotional; deontic (unconditional observance of duties), not utilitarian (aiming at some collective—or even individual—gains in welfare). There is nothing wrong with "generous" acts that elicit a "warm glow" or promise salvation to the giver; these are not to be linguistically conflated with deontic actions, those emanating from pure sense of duty.

the surface but, to the nonsucker, not exactly the same thing.*

In science you need to understand the world; in business you need others to misunderstand it.

I suspect that they put Socrates to death because there is something terribly unattractive, alienating, and nonhuman in thinking with too much clarity.

Education makes the wise slightly wiser, but it makes the fool vastly more dangerous.

The test of originality for an idea is not the absence of one single predecessor but the presence of multiple but incompatible ones.

Modernity's double punishment is to make us both age prematurely and live longer.

An erudite is someone who displays less than he knows; a journalist or consultant, the opposite.

Your brain is most intelligent when you don't instruct it on what to do— something people who take showers discover on occasion.

If your anger decreases with time, you did injustice; if it increases, you suffered injustice.

---

* I need a qualifier here. There are exceptions, but there are also many known cases in which a prostitute falls in love with a client.

PRELUDES

The person you are the most afraid to contradict is yourself.

An idea starts to be interesting when you get scared of taking it to its logical conclusion.

Pharmaceutical companies are better at inventing diseases that match existing drugs, rather than inventing drugs to match existing diseases.

To understand the liberating effect of asceticism, consider that losing all your fortune is much less painful than losing only half of it.

To bankrupt a fool, give him information.

Academia is to knowledge what prostitution is to love; close enough on

ECONOMIC LIFE AND OTHER VERY VULGAR
SUBJECTS 087
THE SAGE, THE WEAK, AND
THE MAGNIFICENT 093
THE IMPLICIT AND THE EXPLICIT 099
ON THE VARIETIES OF LOVE AND NONLOVE 103

THE END 107
POSTFACE 109
ACKNOWLEDGMENTS 115

CONTENTS

PRELUDES V

COUNTER NARRATIVES 001

MATTERS ONTOLOGICAL 007

THE SACRED AND THE PROFANE 011

CHANCE, SUCCESS, HAPPINESS, AND STOICISM 015

CHARMING AND LESS CHARMING SUCKER PROBLEMS 023

THESEUS, OR LIVING THE PALEO LIFE 029

THE REPUBLIC OF LETTERS 035

THE UNIVERSAL AND THE PARTICULAR 043

FOOLED BY RANDOMNESS 047

AESTHETICS 053

ETHICS 057

ROBUSTNESS AND FRAGILITY 063

THE LUDIC FALLACY AND DOMAIN DEPENDENCE 069

EPISTEMOLOGY AND SUBTRACTIVE KNOWLEDGE 073

THE SCANDAL OF PREDICTION 077

BEING A PHILOSOPHER AND MANAGING TO REMAIN ONE 081

Bibliotheca), Procrustes owned two beds, one small, one large; he made short victims lie in the large bed, and the tall victims in the short one.

Every aphorism here is about a Procrustean bed of sorts—we humans, facing limits of knowledge, and things we do not observe, the unseen and the unknown, resolve the tension by squeezing life and the world into crisp commoditized ideas, reductive categories, specific vocabularies, and prepackaged narratives, which, on the occasion, has explosive consequences. Further, we seem unaware of this backward fitting, much like tailors who take great pride in delivering the perfectly fitting suit—but do so by surgically altering the limbs of their customers. For instance, few realize that we are changing the brains of schoolchildren through medication in order to make them adjust to the curriculum, rather than the reverse.

Since aphorisms lose their charm whenever explained, I only hint for now at the central theme of this book—I relegate further discussions to the postface. These are stand-alone compressed thoughts revolving around my main idea of how we deal, and should deal, with what we don't know, matters more deeply discussed in my books *The Black Swan* and *Fooled by Randomness*.[*]

---

[*] My use of the metaphor of the Procrustes bed isn't just about putting something in the wrong box; it's mostly that inverse operation of changing the wrong variable, here the person rather than the bed. Note that every failure of what we call "wisdom" (coupled with technical proficiency) can be reduced to a Procrustean bed situation.

PROCRUSTES

Procrustes, in Greek mythology, was the cruel owner of a small estate in Corydalus in Attica, on the way between Athens and Eleusis, where the mystery rites were performed. Procrustes had a peculiar sense of hospitality: he abducted travelers, provided them with a generous dinner, then invited them to spend the night in a rather special bed. He wanted the bed to fit the traveler to perfection. Those who were too tall had their legs chopped off with a sharp hatchet; those who were too short were stretched (his name was said to be Damastes, or Polyphemon, but he was nicknamed Procrustes, which meant "the stretcher").

In the purest of poetic justice, Procrustes was hoisted by his own petard. One of the travelers happened to be the fearless Theseus, who slayed the Minotaur later in his heroic career. After the customary dinner, Theseus made Procrustes lie in his own bed. Then, to make him fit in it to the customary perfection, he decapitated him. Theseus thus followed Hercules's method of paying back in kind.

In more sinister versions (such as the one in Pseudo-Apollodorus's

TO ALEXANDER N. TALEB

THE BED OF PROCRUSTES

PHILOSOPHICAL AND PRACTICAL APHORISMS

FROM THE AUTHOR OF
THE BLACK SWAN AND ANTIFRAGILE
NASSIM NICHOLAS TALEB